LARGE SCALE COMPOSTING

LARGE SCALE COMPOSTING

M.J. Satriana

NOYES DATA CORPORATION
Park Ridge, New Jersey London, England
1974

Library of Congress Catalog Card Number: 74-82356
ISBN: 0-8155-0549-3
Printed in the United States

Published in the United States of America by
Noyes Data Corporation
Noyes Building, Park Ridge, New Jersey 07656

Foreword

This Pollution Technology Review surveys information available to the beginning of 1974. It is based primarily upon reports produced under the auspices of the Environmental Protection Agency. Several of the studies have been extended abroad, especially to Central Europe, where experience with large scale composting exists.

Large quantities of compost may never be used in agriculture in the United States of America, but as a disposal method for urban refuse it may have merit. Composting is a microbiological process which is capable of converting municipal refuse and other organic waste solids into a product which has a lower bulk than the original waste, is stable, and can be recycled into the ecological system.

In the United States, we are fortunate in receiving direct help from the numerous surveys, together with active research and development programs that are being supported by the Federal Government to help industry and municipalities to control their wastes and rubbish.

In this book are condensed vital data that are scattered and difficult to pull together. Important processes are interpreted and explained by actual case histories. This condensed information will provide a sound background for action towards combating pollution by solid waste management through composting.

Advanced composition and production methods developed by Noyes Data are employed to bring new durably bound books to the reader in a minimum of time. Special techniques are used to close the gap between "manuscript" and "completed book." Industrial technology is progressing so rapidly that time-honored, conventional typesetting, binding and shipping methods are no longer suitable. Delays in the conventional book publishing cycle have been bypassed to provide the user with an effective and convenient means of reviewing up-to-date information in depth.

The Table of Contents is organized in such a way as to serve as a subject index and provides easy access to the information contained in this book.

Contents and Subject Index

Introduction

The handling, processing, and disposal of solid waste have reached such proportions in the United States that solid waste management is one of the major pollution problems today. The citizens of the United States discard over 200 million tons of solid waste annually (about 6 lb/person/day). In the past, most of this waste was burned in open dumps or deposited in landfills. Because of air and water pollution, more satisfactory methods must and are being developed for solid waste disposal.

The real solution to a solid waste problem must envelop the concept of total environmental control. The answer must be a method which will neither contaminate the water, pollute the atmosphere nor desecrate the land. Composting has the potential for providing a rational approach to solving the solid wastes dilemma without polluting the air or water resources.

Composting may be defined as a process in which microorganisms break down organic matter to a relatively stable humus-like material. It is distinguished from digestion or fermentation in that the decomposing material is primarily discrete solids with open pore spaces rather than solids in a liquid or semiliquid environment. Composting processes vary from anaerobic to aerobic and from mesophilic to thermophilic decomposition. However, aerobic thermophilic composting is generally believed to provide the most rapid and complete decomposition of the readily oxidizable or putrescible matter.

Composting is one of the oldest solid waste disposal methods known to man. The process has the capability of converting municipal refuse and other organic waste solids to a product which has a lower bulk volume than the original waste, is stable, and has the potential of being recycled into the ecological sphere without destruction of its high energy value. Compost cannot be considered a fertilizer. Its main value seems to be its high organic content as a soil conditioner, which may provide poor soils with better tilth, water-holding capacity, and improved nutrient-holding capacity. The present technology of composting will permit the recycling of organic waste materials back to the soil without significant pollution of water or land resources. Economically, composting does not compete on a net-cost-per-ton-processed basis with either landfilling or incineration of municipal refuse.

There are possibilities, however, for a more satisfactory program of composting for solid waste disposal. Rough quality compost, cheaply produced, without grinding or fine screening, has real potential for the reclamation of spoiled lands (as from mining), for the prevention of erosion, for reduction of the volume of material going into a landfill, and as a cover material for above-grade landfills.

A further and even more favorable avenue of composting practice would be to consider land as an acceptor of compost rather than compost as a benefit to the land. Rough quality compost might be applied at the maximum assimilable rate (perhaps 100 tons/acre/year) to a piece of land. The land would not be used for crop production, but neither would the land be irreparably changed, as with a landfill or dump. If or when the land becomes needed for subdivisions or agriculture, the compost application could be stopped and the land would recover.

The factors that will influence the future of the composting process as a municipal solid waste management tool are the net costs and benefits of the process, as compared with other waste management processes. As new technology is developed and priorities change on the use of land, water, and air, the cost and usefulness of composting, as well as other solid waste management systems, will be influenced by four factors: cost/ton of solid waste for each alternative processing and disposal system; acceptance of more stringent standards for environmental quality; availability of systems to meet the standards; and public policy decisions requiring beneficial recycling rather than land or sea disposal of wastes.

At present, composting is the only solid waste disposal process that salvages the organic fraction. Since the Solid Waste Disposal Act of 1965 specifically directed that research be carried out on the recycling and reclamation of waste materials, it would appear that the composting process cannot be ignored by the managers of solid waste systems as long as it remains the only known means of reclaiming organic solid wastes.

European Composting Practices

The material below is from EPA PB 205,657 (1969) by M.E. Jensen.

While composting is not the universal answer to solid waste management, it does have a place despite the possibility that this place will not become well established in the United States for many more years. There are a variety of uses for compost, but in general its primary use is to supply humus to the soil; and, as yet, humus cannot be artificially produced.

The German farmer has recognized the need to maintain the marginal organic content of his soil by operating a generalized agricultural system that resulted in materials moving from the soil to crops to animals and back to the soil. The efficiency of the system was such that there was a need to supplement the organic content of the soil by a source outside the system.

Thus it was found that compost could produce results that were comparable to the results obtained by commercial fertilizers applied to mineral-deficient soils. The comparable results created some confusion as to the role of compost in the soil, but studies have now shown that compost is beneficial to the soil as a source of humus and not of minerals.

There is some indication that the reformation now occurring in German agriculture may result in an increasing demand for a ready source of humus as the farms become more mechanized and livestock production becomes a specialized industry.

In the United States the vast land areas available per capita have resulted in single use or low level utilization of the land resources. As a result, the soils are considered young and fertile although suffering from poor management in many areas.

If the future of the United States is close to what is predicted, the day will possibly come when a greatly decreased land area per capita will require the preservation of our lands for intensive agricultural production by transforming municipal wastes into compost or humus for the soil.

EQUIPMENT

Storage Pit

Since composting plants receive only the domestic and some small part of the commercial refuse from a community, these plants generally have a small intake of material, ranging from 40 to 200 tons per day. Thus, the storage bunkers (pits) are designed to receive and hold approximately four collection vehicle loads of refuse.

The storage capacity is adequate to maintain plant operations, but when an operational breakdown occurs, an immediate and readily convenient alternate disposal system has to be available. Most of the bunkers are of a trapezoidic design with the two side walls sloping to a conveyor belt along the bunker bottom.

The sloping walls help move the refuse onto the conveyor belt but create a problem by allowing the material to bridge over the conveyor belt. It can be a very difficult job to break this bridge. The best solution to the problem in existing installations has been to construct a vertical false wall along one side producing one straight and one sloped side.

On new installations, a wider conveyor bottom would be recommended (conveyors six feet wide or more); however, none of the European installations visited had such wide conveyor bottoms.

The Joint United States Public Health Service Tennessee Valley Authority composting plant at Johnson City, Tennessee, uses a wide conveyor, which may be an effective solution to the problem.

Most of the installations on the Continent had dust control equipment for the bunker area, but it was either not used, broken, or considered ineffective in operation.

In plants where the storage bunker was too small, refuse would be discharged on the ramp and pushed in by a tractor as the bunker was emptied. General design called for a steel conveyor belt at the bottom of the bunker to feed a rubber conveyor belt that moved the refuse to the next processing stage.

Refuse Preparation

As the refuse moved from the storage bunker to the first processing step, some salvage and sorting operations were performed. Most of the plants were doing away with picking or sorting because of increasing labor costs and the small income from salvage.

In general, the picking operation was inefficient and greater benefits could have been obtained by better education of the collection crews and the residents. Pickers removed rocks, tires, large pieces of iron, bottles, large sheets of plastic, and other materials considered undesirable for composting.

No paper salvage was performed, and in general, the emphasis seemed to be on plastic removal rather than glass. Salvaged wine bottles are still a source of income, but the thrifty European, the increasing use of plastics, and the one-way bottle make bottle salvaging a negligible source of plant income. The nonsalvageable materials removed were placed in large portable containers for transport to the landfill or incinerator.

An over-band magnet, a drum-head magnet, or both, were standard equipment in all compost plants visited. The drum-head magnet is simply an electromagnetic roller at the end of the conveyor belt.

The magnetic field places the ferrous material in a different trajectory than the other materials leaving the belt. The over-band magnet is an electromagnet directly over the conveyor belt.

An endless belt positioned between the refuse and the magnet sweeps the material out of the magnetic field into a storage container. For greater efficiency two or more magnets might be located at different stations in the process.

The iron removed was baled for storage and sale. German scrap iron suffers the same low price as United States scrap and was sold for whatever price it would bring, or else given away in exchange for the removal.

Processing

Grinders: Following intake, the refuse could be processed by grinding, rasping, or the Dano drum, depending on the composting system being used. If grinding is used, commercial grinders are supplied by the Buhler Company (double or single rotor grinders), the Hazamag Company (double rotor grinder), or the Gondard Company.

Gondard grinders are coming into greater use. The Gondard mill uses a 140 horsepower motor to drive a single rotor, having four rows of 12 hammers, at a speed of about 1,000 revolutions per minute.

Refuse is fed into the mill at a variable distance above the rotor, and as the material falls on the rotor, hard, resilient and heavy objects are propelled into a vertical trajectory of about 20 feet to a point where a deflection plate directs the material into a storage compartment. Soft, inelastic material is beaten through the mill grate by the rotor.

The ballistic separation action by this mill actually provides a mechanical sorting or removal of much of the noncompostable material. The Hazamag grinder uses a high-speed double rotor with swinging arms made of flat steel. The Buhler single rotor grinder uses flat steel arms equipped with a pivoting sledge hammer type head at the end. In either grinder, after some wear, the grinding surface can be reversed for additional use. The double rotor Hazamag and the single rotor Buhler mill are used for primary grinding; the double rotor Buhler is generally used for secondary grinding.

As is standard practice for as much European equipment as possible, the grinders are powered by electric motors. It was noted that both the Buhler and Hazamag grinders were considerably overpowered, but the Buhler mill seemed to be equipped with more excess power than the Hazamag. The excess power is required for starting and for overcoming slugs of material, or materials difficult to grind. Hazamag seemed to be relying on the inertia and grinding action of the double rotor to overcome the hard spots. In addition, the Hazamag mill was more heavily constructed than the Buhler. In the double rotor Buhler mill used for secondary grinding, the two rotors turn at different speeds and it was reported that the hammers of the faster rotor had about one-third the life of the slower rotor hammers.

All the grinders are very noisy in operation. None of the mills could grind nylon, especially nylon stockings. The stockings would sometimes accumulate on the rotor until they had to be slashed or burned away with a torch.

Rasp: Developed by the Vuilafvoer Maatschappij (VAM) firm, the rasp is marketed by the Dorr-Oliver Company. It is a vertical cylinder with a diameter of 15 feet and a total height of 7 feet, containing both a grinding floor and a discharge floor. The grinding floor is formed from wedge shaped sections with alternate sections being constructed with round holes (from about 0.5 to 1.5 inch diameters) or with 0.5 inch steel pins projecting from the section.

The pins also extend part way up the wall all around the inside. There are eight heavy hinged arms that can rotate at various speeds up to 25 rpm. The arms move the material over the floor that serves to alternately screen and rasp the material. The arms can lift up over ungrindable material until such time as a trap door can be opened and the rejects swept out the door by the arms.

The material passing through the grinding floor falls on the discharge floor and is moved out a discharge opening by a rotating arm attached to the center spindle. Compared to a grinder, the rasp is physically larger, requires about one-fourth the power, is somewhat quieter in operation, requires less maintenance, and has a higher first cost. The grinding and rasp system of composting appears to be quite competitive in the open market.

Dano Drum: The Dano drum, usually 9 to 12 feet in diameter and 60 feet or more in length, is rotated at a speed of about 1 rpm during the working day and at about ¼ to ½ rpm at night. The raw refuse is usually given magnetic separation before being discharged into the drum, where a slight inclination of the drum requires about 2 to 5 days for the material to pass through the drum, detention time being a direct function of the speed of rotation. The drum is operated half full, and the slow rotation provides aeration and the self-grinding action that is effective in the reduction of the raw refuse. When the material reaches the end of the drum, a short section of 1.5 inch diameter screen is used to remove large pieces of reject material.

The rejects are periodically discharged from the drum via a trap door in the screen. A characteristic of the drum compost is the presence of small balls of paper. There is some thought that the small, moist, porous ball provides a better microbiological environment than the material produced by a rasp or grinder. Compared to a grinder, the drum has the same advantages as the rasp.

A satisfactory aeration system has not yet been designed for the drum, and, as a result, many operations have been known to have serious odor problems. A recent solution developed for the odor problem has been the use of underground tile fields. The hot, odorous gases are drawn from the drum and forced into a perforated tile distribution network covered by about a meter of soil or compost.

The cover material does an excellent job of removing the odors, and the technique is reported to enjoy wide success. When compost is used for the filtering cover material, the hot rising gases create a very nice hotbed, and the compost meisters enjoy performing small-scale growing experiments to demonstrate what their product can do.

CURING

A common method for providing compost curing is the compost windrow. Windrow turning is accomplished by the use of an end-loader and some type of elevating conveyor that piles the material in the triangle shaped windrow. The height of the windrows is a function of the amount of compost produced and the storage area available.

At the June 1967 INTAPUC Conference, a model and diagram of a windrow turning machine was on display. The machine looked like a self-propelled stern wheel from a river boat. The wheel was supported by rails on concrete walls on each side of the windrow. As the wheel moved against the windrow, the paddle or cups elevated the compost to an overhead position where it then fell onto a conveyor constructed in the center of the wheel.

The machine is simple enough and appeared to have the capacity that would be required, but would have the disadvantages of an inability to shred the compost and to clean all of the windrow from the ground. No single test has been developed for establishing the frequency of windrow turning. The frequency is usually established by the following variables: the windrow temperature as measured by a thermometer placed in the windrow, windrow or compost odor, and moisture content as evaluated by the compost meister's nose, sense of touch, and experience. Until effective and reliable test procedures are developed, composting, like brewing, will remain an art dependent on the compost meister to determine how much moisture to add to the raw refuse or compost windrow, the turning frequency for the windrow, and when the compost is ready for sale.

PREPARATION FOR SALE

At most composting operations visited, when the compost is judged to be finished, it will be screened once before loading. In some operations a final grinding will precede the screening in order to provide an evenly blended and uniform texture of material. Since glass is one of the most objectionable materials in compost, most effort is directed toward its removal.

In general, the machines used are one of three types. Ballistic separation is one of the most common principles used for glass and other hard particle removal. The first type, known as the ballistic separator, is the largest machine. Compost falling into one end of the machine strikes a high-speed horizontal rotor. The rotor paddles are often short railroad tracks. The paddles or impellers strike the compost and propel it in a horizontal or slightly upward trajectory.

Hard, dense, and resilient particles will obtain a high speed and long trajectory, while the soft, light and inelastic particles travel only a short distance. The trajectory paths are confined in a hozizontal enclosure, the length and partitioning of which can be adjusted according to the need.

The second machine, the secator, is probably the most common device used, mainly because it is cheap and mechanically reliable. Compost falling from the end of a conveyor strikes a shield or bounce plate that directs the material vertically downward onto a revolving drum. The strike plate may direct the material onto the top center of the drum or slightly off center in the direction opposite that of the drum rotation.

The heavy and resilient particles will bounce off the drum into a collection bin, while the light and inelastic materials are carried on the drum under the strike plate and into the storage area. The speed of the drum and the position of the strike plate are set in accordance with operating experience.

A third machine type used is the inclined plate conveyor. Compost falls onto a steel plate conveyor that moves at an angle upward against the falling compost. Heavy and resilient particles bounce down and off the conveyor while the lighter material is carried upward. This method seems to be the least satisfactory of the three.

Another system observed was a vibrating conveyor belt. The heavier particles and glass would settle downward through the material to the belt. Near the end of the belt a scraper was set and could be moved up or down to remove the upper material from the belt, thus literally scraping the compost from the glass and other materials. In general, any of these systems alone would be considered satisfactory. For improved quality the compost would sometimes be exposed to two or more of the preceding systems, or the same system twice.

This, of course, introduced additional expense. The problem with this equipment is that greater glass removal results in considerable loss of compost and an operational balance between a maximum amount of glass removal and a minimum amount of compost rejection must be attempted.

Screening or sieving of compost is universal at composting plants in order to provide a gradation of the final product. There is, of course, a direct relationship between the amount of screening of the compost and the final cost. Most of the screens were the vibrating type, with a few plants using small drum-type screens. The screening operation was one point in the system where the operator could have a direct influence on the amount of compost produced.

The screen openings could be changed to produce compost in accordance with the greatest sales demands. In the situation where there was no market demand and large stockpiles of material were unwanted, the screen size was sometimes drastically reduced; this greatly increased the reject material and cost per ton of compost produced, although to the untrained eye the plant operation appeared normal.

An experimental sieve for glass removal is being tested. The sieve is an adaptation of a coal sieve that operates on the counterflow principle. The compost enters the top and falls onto a fine mesh, inclined, vibrating screen. Air, blown at high velocities upward through the screen, suspends the compost a short distance above the screen while at the same time moving the material up the slope of the screen, and eventually out of the machine.

The heavier particles remain on the screen and are moved downward and off. Initial tests indicate that the machine works satisfactorily except that some amounts of the heavier compost are removed with the rejects. This may be solved with some modification of the machine or additional processing equipment.

The use of a cross-flow air current for large particle removal was also observed. The air current was directed across the compost falling from a conveyor. The installation did seem effective in removing wine corks, bottle caps, some plastic, and bits and pieces of other material, but the effectiveness is directly related to the preceding operations.

PRODUCTION COSTS

The most frequent questions asked at a composting plant always seem to be how much does it cost to produce the compost, and what is the income from the sales? At most plants these data seem readily available, and the computation (expenses less income) to arrive at the unnecessary operational subsidy seems easy. It usually is not. To attempt to relate economic data from one operation or plant to the same type of plant in a different area, or to a different type of plant, presents an even greater problem. The confusion is compounded when the different economies of two countries enter the picture. Among the many factors that create confusion in economic comparisons are types of refuse received, economics of the area surrounding the plant, plant age, plant mechanization, labor costs, salvage operations, remoteness of plant, and marketability of the compost. The reporting of operational cost data may be confused by the inclusion or exclusion of such things as capital costs, income from salvage, ancillary equipment utilized in the process, taxation or organizational support funds, and so on.

It was observed that most of the compost plants were located adjacent to the sewage treatment plants. The sewage sludge was recognized as useful in the composting process, and composting was considered a good method for sludge disposal. The cost accounting system usually included the expense of the equipment needed to transport and incorporate the sludge with the refuse but seldom recognized, by means of charges to the sewage treatment plant or something similar, the benefits of using composting for sludge disposal. The economy of composting could present a different picture if the expense for an alternate means of sewage sludge disposal were recognized.

In most of the operations the expenses exceeded the income, but the necessary subsidy was justified as a reasonable expense for the solution of the solid waste problem. Even though the system didn't prove profitable, as long as the plant subsidy was kept low, the operation was considered acceptable. In most cases composting was justified on the basis that incineration capital costs were too great, that there was a lack of room for tipping, or that the watershed laws forbade the deposition of any type of raw refuse on the ground.

The law, however, was unclear on the difference between raw and treated refuse. There are no standards or quality control regulations established for compost produced unless the compost is mistakenly marketed as fertilizer. There are definite quality control standards for fertilizer production.

COMPOST SALES

Present technology is such that there is no problem connected with producing compost; the problem is what to do after the compost is produced. One of the most effective organizations for the utilization of compost is Vuilafvoer Maatschappij (VAM) of Amsterdam, Holland. One of the largest stockholders in the company is the Dutch government. The company has three major sections: technique, agriculture research, and sales; the operation of one section directly affects the other sections.

The history of VAM illustrates the evolution that has taken place in composting. VAM's first plant was built at Wijster in 1932 when there was a large demand for coarse, low-quality compost to use in land reclamation. At that time about 80% of the plant's production went to the land reclamation companies.

By 1940, 90% of the plant's compost went to farms and only 10% to the reclamation companies. This drastic change resulted from the fact that during World War II commercial fertilizers were almost impossible to obtain, and it was necessary to utilize compost to maintain much needed crop production. The movement to utilize compost carried through to the middle 1950's, and by then the widespread use of compost had resulted in the recognition of several new uses. Cattle were found to prefer grazing on pasture that had been covered with compost rather than stable manure; mixing compost with farm manure increased the organic content of soil; compost could be used as bedding material for chickens and pigs; it was found that compost was effective in curing anemia in pigs. These uses, of course, increased the demand for compost. In the 20 years between the middle of the 1930's to the 1950's only one composting plant was in operation, but since then a total of 15 additional plants have begun to produce compost.

During the 1950's, vegetable growing, flower cultivation under glass, and flower bulb cultivation became a large horticultural industry in Holland. This new industry had interesting effects on compost production. A finer quality product was required and the processing of the compost became more complex and expensive. This requirement and ability to produce a finer quality product opened up a new market, a market for what is known as refuse heating manure, freshly ground refuse devoid of any sharp particles.

This is in great demand by greenhouses. The essential feature of the material is the large amount of heat produced during the decomposition of the organic matter. It is difficult to try to match the supply and demand aspects of the product because of the large seasonal demand and the need for the material to be fresh.

Flower bulb cultivation along the sandy North Sea coast created a big compost market when it was found that the use of hand-placed reeds and straw could be replaced by mechanized application of compost. The compost resulted in labor savings and, in addition, had the benefit that it could be left and worked into the soil to become additional organic matter. The ever-increasing population density and demand for recreation areas and parks have created a market for a peat-compost mixture. The peat compost in a 1:1 mixture is said to make the land ready for cultivation quicker and to give the plantings a faster start.

Finally, the changing economy of Holland is resulting in more leisure hours for the city dwellers; hours spent growing flowers in window boxes or on small plots of land. It is more agreeable to apply the compost to the soil than to use animal manures, as there is no relationship between the power of a material and its odor.

Thus, while most compost is sold in bulk, there is an increasing demand for small quantities in attractively packaged bags. It is interesting to note that while the compost meister curses plastic as one of the biggest problems in producing quality compost, the sales people find it the most attractive way to market the product.

From the standpoint of total sales, horticultural and recreational uses rank at top, while general agriculture is near the bottom. As a result of the above, the sales demands are met by compost of one of the following classes: horticulture quality, superquality, peat compost, and town refuse heating manure.

The VAM was established to develop and maintain a market for compost. The government is a strong partner in this organization because it recognized that the proper sale of compost provides a solution to municipal manure (refuse and sewage sludge) disposal problems, as well as maintaining the basic fertility of its limited agricultural land.

From its long experience, the VAM company sets forth the following points as being necessary for continued compost sales: (1) a good outlet in the neighborhood of compost producing plants; (2) absence of competitive products such as farm manure, heating manure, etc., unless they have long supply distance; (3) cooperation by a large and well stocked government information center; (4) the production of exclusive quality products that can be guaranteed completely; (5) favorable trading results with the buyers and no government restrictions; (6) very dry years when crops have had bad yields or died early because of lack of organic matter; (7) the importance of marketing a product with a trademark.

In conclusion, VAM does not hesitate to state that the production of compost is not a money-making proposition. The government is, of course, providing support for the organization, but the amount of the support is undefinable. The important fact is that the compost is sold.

SPECIAL SYSTEMS

A composting system can be considered to have two parts; one part is for the preparation of the material and the other is for providing the required biological environment. Most of the present composting systems can be placed in one of four categories: (1) composting in fermentation cells with prior crushing of the refuse, followed by windrow curing; (2) composting in fermentation cells, followed by windrow curing; (3) composting in windrows without prior preparation, followed by compost product preparation; (4) composting in windrows with prior treatment of refuse without subsequent treatment of compost. The following is a discussion of several special composting operations observed. While the systems presented are unique and generally the only operations of the kind in the world, the system components still fit the above categories.

Van Maanen

The operation, first established in 1932, could be nicknamed the open-dump composter. The majority of the refuse received is railhauled from Den Haag, Holland, about 175 kilometers to the composting plant at Wijster, Holland. The rail cars simply discharge the refuse onto the ground from rail spurs built on viaducts. Immediately after discharge, water sprinkling systems are operated to moisten the refuse.

The operations are scheduled to provide up to 7 months storage in the windrow if possible. After the desired curing time, the compost is transferred via gantry cranes and an endless train of mining cars to the processing building. The material brought to the building is first subjected to a thorough screening by successively coarser screens. The fine material first removed by the screen and given a magnetic separation is nearly free of glass and other undesirable items and makes the best quality compost produced. The material removed by the coarse screens is treated by magnetic separation, then either by a roller crusher or a hammermill. This lower grade compost can then be further refined by ballistic separation and stockpiled for sale. Rejects are used to raise the level of the property adjacent to the windrow area.

The operation is well isolated from nearby farms. As would be expected there are vector and odor problems and a potential groundwater pollution problem connected with the operation. The Dutch government uses the 60 acre site as an experimental laboratory for evaluating vector control programs. Odors are a problem and efforts to use chemical masking agents have not been successful.

An underdrain system designed to collect the leachate has been destroyed by corrosion, and some work is now being done to develop a new collection system and provide some measure of treatment for the water collected. A barge canal near the site was used to transport some of the compost to market, but trucks have now replaced the barges. It was reported that the fish in the canal were used as biological monitors for water pollution problems. This is undoubtedly the lowest cost composting system in operation, but not necessarily the most desirable.

Baden-Baden

The system was developed and is used at the city of Baden-Baden, Germany. It might be nicknamed the modified tip composter. While being somewhat similar to the Van Maanen system, the operation is on a much smaller scale and provides better environmental control. Refuse is discharged from the collection vehicles into a small storage bunker, then lifted by crane to an overhead receiving hopper.

The hopper feeds a rotating drum that was formerly used to screen out ashes but now serves as a flow equalizer for a sorting conveyor where magnetic separation and some hand-picking is performed. The refuse passes from the conveyor into a second drum and is mixed with sewage sludge (from sludge drying beds) until a 50 to 60% moisture content is developed, and then the material is discharged into a truck or wagon for transport to the windrow area.

The windrows are constructed on two lines of large diameter, perforated concrete pipe. In former times air was blown into the pipes and out through the windrow, but odors became a very serious problem. The odor problem has been corrected by drawing air into the pipes through the windrow; the air drawn in is used as the air supply in the bulky waste burner on the site.

Water is applied to the windrows and they are turned if necessary. Water pollution problems are handled by discharging drainage into storm drains or to the surrounding marshy area. After an appropriate period of curing, the compost is trucked to the plant, ground, and sieved for sale.

The primary purpose of the plant is reported to be the satisfactory disposal of sewage sludge and not the production of compost. The compost that is not sold is used to raise the level of an adjacent marsh.

Caspari-Meyer Brikollare

This process could be nicknamed the cookie-cutter composter. The Brikollare system is probably one of the more highly mechanized and expensive systems developed. The system actually replaces windrow curing with a block or controlled environment curing.

Refuse delivered to the plant is processed by magnetic separation and some hand sorting before discharge into a Dorr-Oliver rasp. After the rasp, the material is again processed by magnetic separation and ballistic separation. The material is then transferred via conveyor to the brick machine. Raw sludge, dewatered by a vacuum filter, is conveyed to a pug mill blending tank where three parts refuse and one part sludge is blended to produce a mixture with about 50 to 60% moisture content.

After blending, the material is conveyed to a holding or surge tank for the bricking machine. By an elaborate and complex mechanical system, a measured quantity of the material is placed in a mold, vibrated for compaction, extruded from the mold, and stacked on a wooden pallet. The resulting brick, measuring about 15 x 15 x 20 cm, has the appearance of dark, raw papier-mache.

A half moon slot parallel to the long axis, runs along the middle of the bottom of the brick. When the pallet is full, a fork lift is used to transport the material to the curing shed. While the bricks are stored for about two weeks in the shed, temperatures of about 60°C are obtained and maintained, and large quantities of mold form. No odors were reported to be produced by the bricks.

The curing process was reported as follows. The group of stacked bricks provides a body of heat that maintains the temperature near the optimum for the desired biological activity. Much of the exterior surface of the brick is open to the atmosphere and aerobic decomposition takes place, while inside the brick, anaerobic activity takes place.

As the products or by-products of the anaerobic activity are transferred outward, they encounter the anaerobic-aerobic interface where they are transformed or modified into unobjectionable products. As the curing continues, the brick moisture moves outward and air moves inward and the microorganisms maintain the balance.

Experiments at the plant have shown that when the ground-up bricks are used as insulation material around the brick stockpiles, the stored bricks can still obtain air, and the stockpiles can be stored almost indefinitely. The bricks have the advantages of no odor in long-term storage and a reduction in the size of storage area needed. When the bricks are to be sold, they are passed through a hammermill and the compost is then screened to the desired grade.

It is necessary to process only as many bricks as may be needed to produce the quantity of compost desired by the buyer. Although this system is expensive and is vulnerable to mechanical breakdowns, it provides excellent environmental control and produces a desirable product.

Humboldt

The Humboldt system is experimental and is now being developed and tested. The refuse is processed by magnetic separation while being conveyed from the storage bunker to a pug mill for shredding and mixing with sewage sludge. The pug mill has been modified to include shearing knives spaced at intervals along the twin rotors. The homogenized and shredded material discharged from the pug mill is placed in a tall, conical drum that looks very much like an elongated concrete mixer.

The material moves downward by gravity and is discharged at the bottom by a scraper system. At the center of the drum are pipes used for the injection of air and moisture when necessary. Hopefully, the system will not require windrow curing and has the advantages of a minimum of mechanization, reduced power requirements, and a controlled environment.

Nylon stockings present a problem in the pug mill because they do not shear and then become wrapped about the rotor and entangle pieces of refuse. The result is a huge ball of material that jams the machine. The process is patented and now operating on a pilot scale.

RESEARCH

Soil Effects

Studies carried out indicate that compost has definite short and long-term advantages when added to the soil for crop production. The exact mechanisms that create these benefits are unknown, but seem to be tied into the mysteries that still surround the material known as humus. Compost has been shown to increase the biological activity in the soil; this has been related to the fact that compost produced crops experience less plant disease. The compost mineral analysis is well-known and reveals that the material is definitely not a fertilizer; yet when applied to the soil, the compost produces a crop response that is very similar to that obtained with commercial fertilizer.

Compost has been proven to provide a rapid start for plant growth and seems to be able to sustain the quick start advantage through to the maturity of the plant. One hypothesis is that the dark color of the compost provides a better retention of radiant heat, and the radiant heat aids the growth process. Dr. Koepf at the Agricultural University at Stuttgart-Mohringen is undertaking studies on the relationship of nitrogen and the soil. Dr. Koepf and his colleagues have just completed work on the background levels of nitrogen present in natural streams, and, in an effort to understand the source of the background levels, Dr. Koepf plans to study the fate of nitrogen applied to the soil. The studies may indicate how much nitrogen goes to the atmosphere, how much nitrogen enters the soil; and of the nitrogen that enters the soil, how much passes through the soil to the water, how much is utilized by plants and microorganisms, how much is retained by the soil, and what forms of nitrogen can best be utilized by the life systems of the soil.

It was noted that when immature compost was being sold, the compost meister usually advised the buyer not to use the material for at least two weeks. The meister had learned by experience that early application of the compost to soil results in adverse effects on crops because the microorganisms extracted nitrogen from the soil in order to complete their actions on the compost.

A test for compost maturity in relation to nitrogen was observed. The test apparatus is also used for determining soil moisture in situ and consists of a pair of tongs with a wedge-shaped head that clamps about a special piece of porous paper. The instrument is inserted in the soil for a period of time, removed, and a measurement of the penetration of the moisture from the paper edge is referenced to a correlation curve to determine soil moisture. The same procedure is used in a compost sample with specially treated paper.

The moisture content of the compost is measured as above, and then chemical additives are placed on the moist area. Color change determines absence or presence of nitrogen, with color intensity providing indication of the quantity. This test can be correlated with the compost maturity and used for guidance in future operations. Test accuracy was not guaranteed and its mechanism was not completely understood.

Erosion Control

The high population density of Germany requires maximum production from the agricultural land available, and sustained high level productivity over a long period of time, so that soil losses must be minimized. One of the prominent German crops with a tremendous erosion potential is grapes grown on hillside vineyards. The grapevine growth is such that there is no ground cover, leaving the soil with a maximum exposure to the erosion-producing elements of nature.

In the past, the hillsides were terraced for erosion control, but this is not an efficient utilization of the area available and gradually the terraces have been replaced by the open mountainside. Both the German government and the Wine Growers Association have recognized the potential problem and have put forth money to determine what control techniques can be utilized.

Compost has been found to be a very effective material for erosion control, and the wine growers are probably the largest group using compost in Germany. Dr. Bosse, at the Bad Kreuznach compost plant, is employed by the Wine Growers Association to study the relationship of compost to erosion control and soil improvement in vineyards. About 90% of the plant production was being used on vineyards. On the hillsides where hand labor is necessary, the vineyard rows are planted 1.2 meters apart, while on the flatter areas the rows are 3.2 meters apart to allow the use of machinery. Studies have found that while the method of plowing has some influence, the application of 100 cubic meters of compost per hectare every three years is effective in erosion control. Based on the fact that compost has been demonstrated to be an excellent material for erosion control, studies were in progress to determine the optimum amount and frequency of application needed to provide long-term benefits and maximum crop production by soil improvement with compost.

An interesting erosion control study has been made by German scientists. The question arose as to how much soil had been lost by erosion over the years. Hundreds of years ago when the land was first being cultivated, the hillside rocks were quite a nuisance. In clearing the fields the rocks were piled along boundary lines or at whatever distance a man could stagger with a large rock or throw a small one.

The results of these efforts can be seen today as long rockpiles along the hill slopes. It was agreed that the earth under these piles should be at approximately the original elevation and, therefore, some of the windrows of stone were moved and precise measurements were made of the differences in elevation of the original and the present ground levels. The results indicate that approximately 3 inches of soil had been lost in several hundred years.

Pathogen Studies

Detailed studies have been completed in Germany on the subject of pathogen survival during the composting process. The principal investigator for these studies was Dr. D. Strauch, a doctor of veterinary medicine at the Justus-Liebig University Institute of Human Health Protection and Animal Health at Giessen, Germany. His primary interest was the epidemiologic study of the role of composting in relation to man, animals, and the soil.

His studies were performed on a variety of compost processes, and with a high degree of probability concluded that as long as proper temperatures, humidity, gas exchange, and detention times were maintained in the process, the end product contained no hazards for man or animals.

Dr. Stauch developed several interesting techniques during his studies. In one technique he used a Perlon plastic or cloth mesh bag that he named the inoculation bag. Two of these bags usually were used in his studies. Both bags were filled with 100 grams compost or ground refuse, and a hypodermic syringe or gelatin capsule was used to inoculate the bag with a known number of the organisms to be studied; a sealed ampule containing a known quantity of the same organism was placed in the other bag.

The two sealed bags were then placed at a desired location in the compost windrow. The ampule provided data as to whether the pathogen destruction was a temperature function alone, while the bag, allowing free exchange of gases, nutrients, and most everything except the test organism, provided data on which microbiological interactions might be operating to destroy the pathogens. The results obtained were considered highly satisfactory. A different technique was required when the composting process included a fermentation cell such as the Dano drum. For this Dr. Strauch developed what he called his sputniks; hollow, perforated metal spheres that were bolted together. He would place his inoculated bags in these spheres and expose the spheres to the composting process.

In the first attempts to use the spheres in the Dano drum, it was found that they were quickly bounced through the drum. In order to closely simulate the movement of the material through the drum the spheres were wired into place inside the drum and periodically the drum would be entered and the spheres transferred to a new position. Dr. Strauch felt the results obtained were quite reliable. Among the organisms tested by Dr. Strauch were *Bacillus anthrax, Salmonella enteritidis, Erysipelothrix rhusiopathise* and Psittaservirus.

Research work on the other aspects of composting had found that the optimum temperature for biological activity in composting is in the range of 40° to 50°C. With a humidity between 40 to 60% even the most resistant sporozoa are killed. It was found that antibiotic substances produced under the above conditions operated to control the pathogenic organisms.

In developing controls for the production of hygienic compost, two points must be kept in mind. The necessity for an optimum environment is pointed out above. The second point relates to how compost improves the soil. It is known that the processes occuring in compost enhance the soil biological activity, that needed plant micro- and macronutrients are produced, and that the compost finally becomes humus.

The mysteries of humus are still unsolved. In order to produce the preceding benefits, compost must be placed on the soil as soon as possible. Therefore, as soon as the decomposition of the refuse in an optimum biological environment has satisfied the public health requirements, the process should be terminated in order that beneficial continuing actions can take place in the soil.

European Activity and American Potential

Solid waste management in Europe has long been lauded as a model of efficiency and appropriateness. The argument or explanation has been that, in densely populated Europe, one simply must be right about waste management, and the Germans, Swiss, Dutch, and other Europeans are right.

One of the foremost claims has been of the success of composting, and two examples are always cited to substantiate the claim. The first is that nine composting plants have begun operation in Germany since World War II. The second is the success of composting manufacture and sale by the VAM or Waste Transporting Corp., Amsterdam and Wijster, Holland. That corporation makes 160,000 of the 200,000 tons of municipal compost produced in that country each year, and some eight cities and towns, including the capital of Den Haag (1 million inhabitants), deliver all their municipal waste to that organization.

These examples and others do point up the success of these European composting activities. Nevertheless, even these successes are not unqualified (as will be discussed later), and their success cannot be transferred automatically and unequivocally to the American scene. Still, an indepth study of the situation in Europe may assist understanding what might be appropriate in the United States.

A study of European composting activity is described in EPA PB 205,656 (1968) prepared by S.A. Hart of the University of California. This study was primarily concerned with utilization of compost after manufacture. In the pursuit of this goal, 14 European composting plants were visited. In addition to learning how these plants dispose of their compost, information was obtained on their composting equipment, facilities, and procedures.

Composting is more applicable to a city's domestic refuse than to commercial or industrial refuse, and composting is not generally appropriate to debris from construction and demolition. Thus, most European composting plants accept only domestic refuse and have alternate or supplementary facilities for handling the noncompostable parts of commercial and industrial refuse. Even with domestic refuse, a considerable percentage (15 to 30% by weight) is not compostable and must be removed before the compost can be sold or used. The details of this process and others are best explained as they apply to individual plants.

SURVEY OF FOURTEEN EUROPEAN COMPOSTING PLANTS

Germany

Bad Kreuznach: Population Served 45,000; Type of Plant, Dano – One of the most successful composting operations in Europe is at Bad Kreuznach. This city, near the center of the Riesling wine area, is on the Nahe River, a tributary of the Rhine. Much of the vineyard acreage of the area is on steep hillsides, where added organic matter is needed to prevent excessive soil erosion.

Manures and similar organic materials have been in short supply in the Bad Kreuznach area. Thus, in 1958 the BVB (Bodenverbesserungverband, or Soil Improvement Cooperative) of the local grape farmers (under the leadership of its president, Mr. O. Andres) entered an agreement with the city and the state. The compost plant, belonging to the farmer cooperative, was built on city land adjacent to the sewage treatment plant. The city's public works department collects and hauls domestic refuse to the compost plant, where it is discharged into the composting drum.

The city pays DM 80,000 per year for this dumping privilege, which is less than the cost of hauling the waste to a more distant landfill and burying it. (The landfill is used for commercial and industrial wastes, except winery wastes.) Sewage sludge from the adjacent plant can be and has been incorporated into the compost, but at present it is hauled away as liquid sludge by the farmers and used in that form.

The unground, raw domestic refuse brought to the composting plant, along with raw winery wastes in winter (the lees, or skins, seeds, and stems), is put into the Dano drum. Residence time there is 3 to 4 days. The partially composted material is then removed from the drum. Iron is removed with a magnet, and the compost is sieved.

The compost is then piled into windrows for curing and storage until purchased by the farmer, in the fall of the year. Thus, some compost is stored 9 or 10 months. The noncompostable material, scalpings from the sieve, are used for fill material at the compost site. (The plant was located on low-lying ground, and the approximately 10 acre compost storage site has been raised about 6 feet in 10 years of operation.)

About 8,000 cubic meters (10,500 cubic yards) of compost are manufactured each year from the incoming 25,000 cubic meters (33,000 cubic yards) of raw domestic refuse. The compost is sold to the Cooperative members for DM 10 per cubic meter ($1.90 per cubic yard), f.o.b. the compost plant. This revenue of DM 80,000 ($20,000), plus the city's payment of DM 80,000, just equals the cost of making the compost (labor, maintenance, and equipment amortization).

The state's contribution to this venture has been to furnish a part of the capital and to pay the salary of a research scientist attached to the operation. These research scientists have done research on the composting operation itself and on compost utilization. The compost utilization research has consisted of measuring runoff water and runoff soil from some steep experimental plots.

The success of the Bad Kreuznach composting operation is due to all the above factors, economy to the city, state support, and, most important, a good market for the compost. All factors may well be required for similar American success.

Blaubeuren: Population Served, 20,000; Type of Plant, Windrowing of Shredded Refuse – Blaubeuren is a small community in the Alb Mountains of Germany. A large cement plant in the city uses the calcareous rock of the surrounding area as raw material for cement manufacture. Dr. E. Spohn, mill manager and part owner, has long believed in returning the ore stripped land to agricultural productivity and has advocated and been demonstrating the value of compost from municipal waste for this reclamation.

Blaubeuren does not produce enough compost to reclaim the land at the rate it is being consumed, composting is rather expensive for such a small community, and no tangible, direct measure of the value of land reclamation is yet available. Not only is Dr. Spohn personally knowledgeable in composting technology, but he has equipped his cement manufacturing operation with a research laboratory and an experimental farm for the purpose of studying this special field of compost utilization.

The Blaubeuren compost plant itself is small and unpretentious. The incoming raw domestic refuse is run through a Dorr-Oliver rasper. Burnable uncompostable refuse is disposed of in a simple incinerator. The raw rasped refuse is piled into windrows and cured, with one or two turnings in 3 to 6 months. Most of it is then used in the land reclamation research discussed above. Sewage sludge is sometimes added to the ground refuse before it is put into windrows, but, since the sewage treatment plant is some distance away, the sludge (cake or slurry) must be trucked in.

Duisburg-Huckingen: Population Served, 90,000; Type of Plant, Dano – Duisburg is a major industrial city on the western end of the Ruhr River. The city had been having trouble with sewage sludge disposal and decided that composting domestic refuse plus sludge could be economical and satisfactory. Therefore, a two-drum Dano plant was built at one of the sewage treatment plants, and refuse from 90,000 of the 400,000 inhabitants is converted into compost. Incoming raw refuse is handpicked for glass and rag salvage, and iron is removed by a magnet. The refuse, otherwise untreated, is then put into the 11 ft diameter 60 ft long Dano drums, along with sewage sludge, and tumbled for 3 to 5 days. The fresh compost is sieved and piled outside to cure.

The noncompostable sieve scalpings are buried. This plant has had trouble with odors from the composting operation. The plant, once in an undeveloped area, is now surrounded by upper middle class apartments. At one time, odors from composting (anaerobic conditions within the Dano drum) were so objectionable that the plant had to be shut down in summer when the incoming refuse was wettest. (In winter there are fewer vegetable and garden trimmings, and the ash content is higher.)

The odor problem has been solved by a combination of techniques. First, the sewage sludge is thickened or partially dried to add a minimum of moisture to the system. Even then, only one-third of the population equivalent of sludge can be added in summer, and one-half the population equivalent is added in winter. In addition, all the exhaust ventilation air of the building and of the Dano drums is scrubbed through a soil filter. This filter consists of a buried perforated pipe covered with earth and cured compost; the filter, approximately 10,000 square feet in size, filters about 7,000 cubic feet of air per minute.

The cured compost is usually rescreened, and may be ground or crushed and sieved again, resulting in a fine textured, homogeneous, acceptable compost. It is sold primarily for gardening, nursery, and landscaping use, with none going to general agriculture. 8,000 to 10,000 metric tons (9,000 to 11,000 short tons) are made per year. Over three-quarters of Duisburg's domestic refuse, plus all the commercial and industrial refuse, is handled other than by composting.

Heidelberg: Population Served 30,000; Type of Plant, Multi-Bactor Compost Tower – Heidelberg is the world famous university city in southwest Germany. Several years ago city engineer O. Horstmann began investigating composting as a method of processing the city's domestic refuse. The first idea was to sort, rasp, and ballistically separate the refuse and then place it in large windrows (about 35 ft wide and 20 ft high). Anaerobic conditions within the windrows caused foul odors, and it was difficult to turn the windrows. In 1962, therefore, a mechanical composting tower, the Earp-Thomas Multi-Bactor unit was installed and is working satisfactorily. Today this unit is making about 2,000 cubic meters (2,600 cubic yards) of compost per year.

Sewage sludge is added to the Door-Oliver rasped refuse and conveyed to the top of the eight stage 20 ft diameter 35 ft high composting tower. A vertical shaft turns plows on each stage. The refuse is thus stirred continually, and it gradually drops from stage to stage through holes in each stage. The shaft and plows operate 8 hours per day, and the refuse residence time within the tower is 3 days.

The removed compost is sieved and dropped onto an impact separator known as a secator. Glass, slag, and hard objects bounce one way, while the high quality and softer compost rolls the opposite way. The compost is stockpiled temporarily, until it is taken away by the wholesaler, but longtime curing does not appear to be necessary. The wholesaler arranges to sell and transport the compost and sometimes even to apply it. Essentially all of the compost is used for home gardens and landscaping projects.

The wholesaler does dispose of all the 2,000 cubic meters (2,600 cubic yards) per year, but that appears to be the limit of the market although Heidelberg could make more compost by processing more of the city's domestic refuse.

Schweinfurt: Population Served, 85,000; Type of Plant, Caspari-Brikollare – The Brikollare Verfahren, or briquette process, is used to make compost at a plant erected in Schweinfurt in 1965. Previously, all municipal solid wastes, including industrial waste from two large ball-bearing manufacturers and one heavy machinery fabricator, were buried in a landfill. That was changed with the construction of a $1,000,000 composting plant for domestic and commercial waste and a $1,250,000 incinerator for industrial waste.

The incoming domestic and commercial refuse is elevated from the receiving bunker by a belt conveyor. One man inspects the refuse as it is conveyed and removes materials that might damage the mechanisms, but he does not really salvage any material. Iron is removed by a magnetic separator, and then the refuse falls into Dorr-Oliver raspers for shredding. The shredded refuse is then run through a ballistic separator, where gravel, broken glass, and similar hard, heavy material is removed. The ground compostable refuse is then elevated to a surge tank.

Meanwhile, digested sewage sludge from the city's treatment plant is vacuum filtered to increase the solids content from 12 to 30% (nearly three-quarters of the water in the sludge is removed). The dewatered sludge is mixed with the ground refuse in an auger conveyor and conveyed to the briquetting machine. Here, an elaborate machine produces briquettes, approximately 15 x 9 x 6 inches in size, that have a tunnel on the underside.

These are placed automatically on pallets that are moved to a curing shed, where the actual composting takes place. The briquettes quickly heat to 130° to 140°F (55° to 60°C), a surface fungal growth develops, and the briquettes both dry out and compost. The salient feature of the process is this concurrent biological-physical change. The metabolic heat of composting promotes surface drying, while the compacted nature of the refuse caused by the pressing of the briquettes allows capillary transfer of moisture from the center to the surface.

Concurrently, as moisture is removed, air enters the capillaries of the briquette and sustains the aerobic fungal and bacterial organisms that attack the refuse and convert it to compost. The close stacking of the individual briquettes, along with their tunnel form, gives adequate opportunity for air transfer while still conserving the heat of composting.

Even the outside corner briquettes attain a temperature of 120° to 130°F (50° to 55°C). In the curing process the moisture content drops from about 65 to 13% (wet weight basis). The cured briquettes are stacked outside in a yard like normal building bricks. In the fall of the year, when compost can be marketed to nearby grape farmers, the briquettes are run through a simple hammermill, and the finished compost is sold. The impressive features of this Caspari process are the odor-free operation of the biological-physical composting process and the minimal land area that is needed. Additionally, the vacuum filtration of the sewage sludge makes possible the disposal of both domestic refuse and sewage sludge. Some 10,000 metric tons (11,000 short tons) of compost are produced each year.

Compost made by this process is of good quality, but probably is no better or worse than compost made by any other process. The quality of compost depends upon how much grinding, screening, sieving, ballistic separating, and similar physical upgrading are done. This upgrading can be done after composting, as in the Dano process, or before, as in the Caspari. Probably neither process offers any great economy over the other. The Schweinfurt operation appears to be quite economical of labor, but amortization of the large capital investment is a significant cost.

St. Georgen: Population Served 14,000; Type of Plant, Windrowing of Ground Refuse – St. Georgen is a small spa, or resort town, in the Black Forest, southwest of Stuttgart. The city's small compost plant takes the community's domestic and commercial waste (there is no industry other than tourist) plus sewage sludge, and converts it into compost. The refuse is handpicked for salvageable items, iron is removed by a magnetic separator, and the refuse is ground with a Buhler hammermill.

Noncompostable material (mostly commercial refuse) is burned in a 1.5 metric ton per hour (1.65 short ton per hour) incinerator, and the ash is returned to the freshly ground refuse. This refuse, plus sewage sludge from the adjacent sewage treatment plant, is arranged in windrows (about 10 ft wide and 4 ft high) and cured for 3 to 6 months, usually with three turnings.

About 1,400 cubic meters (1,800 cubic yards) of finished compost are produced each year, and sold to home gardeners, truck crop farmers, and forestry nurseries. The net cost to the city was not reported, but would appear to be rather high because of the small volume of waste handled plus the investments in both incinerator and composting equipment. The plant is neat and well run.

Stuttgart: Population Served 75,000; Type of Plant, Windrowing of Ground Refuse – The Stuttgart composting operation is similar in concept to those of Duisberg and Heidelberg. That is, public works officials recognized that composting was a feasible avenue of management for a portion of the city's domestic refuse, though not for all of it (Stuttgart is a city of 650,000). Therefore, a conventional windrow composting plant was built in 1959 in the suburban district of Mohringen.

This plant is the only one in Germany that was not conceived to take sewage sludge also. The incoming refuse is conveyed from the storage bunker over a picking belt. When salvage prices are favorable, glass, cardboard, clean paper, or rags can be picked off. At present only iron is removed with an overband electromagnet. The refuse goes to a Dorr-Oliver rasper and is abraded. It is then transported by truck some 50 to 100 yards to the windrow area, where composting occurs. The material is turned twice within about 3 months.

There has been no high pressure selling program to dispose of the compost. Citizens of Stuttgart may come to the plant and take, at no charge, small amounts of the material for their homes and gardens. Most of the compost, however, is sold to neighboring grape farmers.

Composting is looked upon at Stuttgart as a public service, not only to satisfy homeowner wants but also to demonstrate the principle of conservation. It is recognized and accepted by the city government that composting costs more than alternative waste disposal methods but that it is an appropriate community expenditure, the same as supporting city parks or museums.

France

Versailles: Population Served, 82,000; Type of Plant, Triga, Silo Type – Incoming domestic refuse amounts to some 150 metric tons (165 short tons) per day. The metal is removed with a magnetic separator, and the refuse is then ground with a heavy duty Hazemag hammermill. The ground refuse is rough screened (scalpings go to an incinerator) and then put into the four silos, each with a capacity of 320 metric tons, about 700 cubic meters (350 English tons, 915 cubic yards).

It was unclear whether the daily input is divided among the four silos or whether one silo is used for each day's waste, in rotation. In any event, some air is blown through the silo for aeration, and each silo is equipped with a bottom unloading device. The removed compost is sieved a second time and then piled outside for curing and to await sale.

This plant, privately owned (by the Triga Co.), charges the city of Versailles for domestic waste accepted. The price and the market for the finished compost were not defined. The plant is very attractive and is apparently a well-engineered unit.

Switzerland

Buchs: Population Served, 40,000; Type of Plant, Windrowing of Ground Refuse – The Buchs compost and waste-burning plant in Switzerland serves some 40,000 people in 16 communities in two countries (Switzerland and Lichtenstein). The actual plant is located in the town of Buchs, population 5,000. Domestic refuse is brought to the plant by municipal vehicles of the various communities.

Commercial and industrial refuse and construction debris are hauled to a regional landfill. Both the market for compost and the facilities of the composting plant are not adequate for all the received refuse to be converted to compost, so a substantial fraction is burned. Composting is marketed primarily to grape growers in the area. The incoming refuse is first ground with a Buhler coarse hammermill. Iron is then removed with a magnetic pulley, and the refuse falls into a second hammermill for fine grinding. The refuse is then sieved. (Scalpings are buried in the industrial waste landfill site.)

Sewage sludge can be and is mixed with the sieved, freshly ground refuse, and the material is then placed in windrows, which are usually turned twice during the following 4 months.

Hinwill: Population Served, 100,000; Type of Plant, Dano – Like most Swiss plants, the Hinwill composting plant serves a number of small communities. Hinwill itself has a population of 3,500. Thus, the wastes are primarily of domestic origin, with only a small volume of commercial and industrial refuse needing disposal. The incoming refuse is run over a magnetic pulley and deposited in the Dano drum along with thickened sewage sludge. The drum itself is 11.5 ft in diameter by 90 ft long, and the refuse has a residence time of 2 to 3 days.

The discharged material is run through a coarse and a fine Buhler hammermill, sieved, and then stored outside for final curing. The material not passing through the sieve is burned on site. This incinerator is too small and has created some operational problems. As at the Dano plant in Duisburg, a soil filter system for the ventilation air had been installed and appeared most successful. The compost is sold primarily for grape vineyards.

La Chaux-de-Fonds: Population Served 15,000 (estimated); Type of Plant, Original Dano – This plant is located in the western section of Switzerland, in the city of La Chaux-de-Fonds, population 43,000. The plant is one of the original Dano installations, with a fast-rotating drum followed by a drum screen. The incoming domestic and commercial refuse from about one-third of the city is put into the conventional Dano drum (at La Chaux-de-Fonds it is about 9 ft in diameter and 40 ft long). The drum is rotated at about 2 rpm for a half hour, to abrade, crush and grind the refuse.

The material is then conveyed to a drum screen, called an Egsetor. This unit is 9 ft in diameter and 12 ft long. An outer shell and an inner cage or sieve of pipes rotates; the crushed, ground, fine material falls through the pipes while cans, plastic, and large inert materials are retained on the pipe screen. The two components are periodically removed from the Egsetor to separate piles, the compostable refuse is skip loaded to a windrow, and the inerts are buried. The compost is sold to small truck crop farmers in the area or transported 30 or more miles for home garden sales.

Turgi: Population Served, 70,000, in 10 Communities; Type of Plant, Multi-Bactor Compost Tower – This plant is similar to the Heidelberg Multi-Bactor composting operation except that it is enclosed within a building and uses different makes of auxiliary equipment.

After iron is removed magnetically, the refuse is ground with a Hazemag hammermill and then sieved before being elevated to the Multi-Bactor tower. After residence in the tower, the compost is sieved again and prepared for market. No curing is provided. A considerable volume of the Turgi compost is put into heavy duty plastic sacks for sale to homeowners. It appeared that less than half of the incoming refuse is made into compost, the division being made on the screen after the grinding. The grinding of the raw refuse reduces volume greatly, so that it can be landfilled more easily.

Holland

Arnhem: Population Served, 134,000; Type of Plant, Windrowing of Ground Refuse – Composting is successful in Holland because compost utilization is highly developed. The incoming domestic and compostable commercial and industrial wastes are run over a magnet for iron removal, and then shredded in a Dorr-Oliver rasping machine. The unshredded material from the rasp, plus burnable bulky objects and noncompostable industrial waste, is burned in an on site incinerator (an operation similar to that at Schweinfurt).

The shredded refuse from the rasping machine is placed in concentric windrows by a pivoting belt conveyor. The engineers at Arnhem are experimenting with removing plastic film from the freshly ground refuse by means of an air blast that blows it out while the refuse falls from one belt to another.

Not all the freshly placed refuse is allowed to cure at the compost plant. At some periods of the year nearly all of the freshly ground refuse is bought by specialty farmers in the area and used in hotbeds. Like horse manure of old, this freshly ground refuse undergoes biological stabilization, producing heat and keeping the plant bet warm.

Freshly placed refuse that is not sold immediately is piled into windrows about 4 or 5 feet high. After the initial heating the windrow is turned and piled higher, into a windrow 12 or 15 feet high, and curing is continued. The compost is not sold directly by the City of Arnhem, but by the VAM organization.

Wijster: Population Served, 1 Million Plus; Type of Plant, Van Maanen System – The Wijster compost plant, owned and operated by the VAM, is in the heath area of northwestern Holland, about 90 miles from Den Haag, source of most of the refuse. Composting has been conducted at Wijster since 1927. Domestic and commercial refuse, plus some industrial waste, is transported to Wijster by special railroad cars from Den Haag and the other contracting cities. The railroad cars are unloaded by gravity from viaducts above the composting area. The untreated refuse is wetted down with a sprinkler system and allowed to begin composting.

After 1 or 2 months the refuse is turned with large grab cranes, with two or three turnings being made during the composting period of 6 to 8 months. When the composting is completed, the refuse is conveyed into the processing plant by cable operated railroad cars. At the plant the refuse is sieved, crushed, and resieved. Several grades of compost are manufactured, including some with peat moss added. The noncomposted residue is transported to an open dump disposal site, where it is distributed and compacted. This is not objectionable, since the material is the inert residue from composting. There are some vector problems at the actual compost site: rats and flies are attracted by the freshly placed refuse.

The Wijster area is relatively uninhabited, however, and aside from the vector problem the composting operation as a whole seems very satisfactory.

COMPOST UTILIZATION IN EUROPE

The nine composting plants of West Germany make compost from less than 1% of the domestic refuse generated in that country. The situation is more favorable in Holland, where one-sixth (17%) of the nation's refuse is made into compost. France, Switzerland, and Italy also compost domestic refuse, but the quantity or percentage figure is not available.

All countries could produce more compost, and would produce it if it could be marketed and utilized. At least in Germany, Holland, Switzerland and France, however, expansion of composting operations is slow because there is little additional opportunity to utilize the compost economically. The present emphasis in solid wastes processing in those countries is on incineration, with the production of steam or electricity.

Compost has its major market only in luxury agriculture. There has been a good deal of research on compost utilization in basic agriculture, however, even with promotional research, compost is not being used extensively in basic agriculture. This is regrettable, for compost could be quite cheaply produced for this use. It would not need to be so refined, with fine sieving, milling, and grinding. The reason is that it does not pay. The nutrients in a ton of compost are almost negligible; a typical compost contains only 0.5% nitrogen, 0.4% phosphorus, and 0.2% potassium. Farmers can buy and apply chemical fertilizers more cheaply than they can apply free compost or even livestock manure.

Advocates of compost point out that the important aspect of compost is the organic matter content, not the nutrients. It is the organic matter that improves the physical properties of the soil, so important in reducing or eliminating erosion on the steep vineyard slopes. In general agriculture, however, soil erosion is prevented more economically with contour farming and similar practices than with compost addition. Under today's economic conditions a farmer of corn, cotton, or general row crops just cannot justify the use of compost.

In 1927 the Dutch located the composting plant for the Den Haag refuse 90 miles away, in Wijster, in the middle of the heath lands. The intent was to reclaim this sandy, marginal land with compost. In the early years the compost was distributed to the nearby land, and some improvement was brought about. But farmers today have learned to farm with chemical fertilizers.

Organic material is not an important factor except for erosion control on steep slopes. Today, most of Wijster's compost is backhauled to the populous, intensively cultivated coastal region of Holland and used in luxury agriculture. The actual market for Dutch compost in two recent years is shown in Table 2.1 on the following page.

TABLE 2.1: SALES DISTRIBUTION OF COMPOST FROM MUNICIPAL REFUSE IN HOLLAND

Outlet	1961	1965
	----Percent----	
Forestland improvement	2.4	0.6
Basic agriculture (field and row crops, and for pig litter)	34.4	16.4
Fruit farming	6.5	6.3
Hotbed vegetable farming*	11.6	13.0
Greenhouse vegetable farming	7.9	8.4
Flower and flower bulb production (greenhouse and outdoor)	11.7	17.6
City park, sportfield, and recreational use	25.5	37.7

*Hotbed compost is freshly ground domestic refuse. It is used on the bottom of the hotbed in place of horse manure. The biological process of composting generates heat that makes the hotbed crop grow. Hotbed crops in Holland are cucumbers, melons, and green peppers.

Germany: In Germany too, the marketing of compost is luxury oriented. About 60% of the compost is used to reduce or prevent soil erosion on steep hillside vineyards that produce the grapes for Germany's fine quality wines. Organic matter has always been needed on these hillsides; in former years the farmers used animal manures and crop residues. Today, such organic materials are less available in the immediate area, because of the mechanization and specialization of agriculture, and the substitute is compost from municipal refuse.

Most of the rest of German compost is marketed for home gardens, park and recreation uses, via similar luxury avenues. The average German citizen produces about 500 lb of domestic refuse per year; between 150 and 200 lb of compost can be made from this amount of refuse. Utilization of 150 lb of compost per citizen per year in luxury agriculture, or in all agriculture, seems unlikely. Other markets for compost are limited.

The Blaubeuren compost plant makes 1,500 to 2,000 short tons of compost per year, it is being used for mineland reclamation. It was also indicated that in some years some German compost is used for fruit production (cherries, peaches and fresh grapes) and for market vegetables, but no information was obtained on the actual quantity so used.

Switzerland: The Swiss utilization of compost is similar. Most of it is used on vineyards, but marketability has been poor. Attempts to move into the home garden market have met with some success, but, here too, the volume that can be moved via this avenue is limited. Two of the four visited plants, Buchs and Turgi, have plans to install or are installing refuse incinerators and in the future will produce compost only to the extent that it is readily marketed.

EUROPEAN RESEARCH IN COMPOST MANUFACTURE AND USE

A number of lines of research in composting have been conducted in Germany in recent years. German composting research has been in such diverse fields as engineering technology (Stuttgart), public health and pathogen survival (Giessen), use in strip minefield reclamation (Bonn), use in vineyards (Bad Kreuznach), and use in general agriculture (Braunschweig).

Engineering Technology – Stuttgart

The European leader in the engineering aspects of solid waste management has been Dr. F. Popel, professor of sanitary engineering at the University of Stuttgart. A part of the research effort of his Institute and of his personal work has been in composting technology. This has encompassed both basic studies on biological processes of composting and practical research in designing composting plants for municipalities.

Professor Popel has long felt that the composting process could be substantially accelerated and made more economical if partially composted refuse were recycled. The concept is similar to the activated sludge process of sewage treatment. Laboratory results at Professor Popel's institute have proved the merit of this approach. Actually, the biological process is but one part of producing compost. The materials handling part is equally important, especially with regard to the removal of noncompostables and grinding and fine sieving to meet today's market. The Caspari-Brikollare process of composting at Schweinfurt is an outgrowth of earlier research on materials handling by Professor Popel. It appears that the more critical part of the composting problem today is utilization.

Public Health Problems with Compost – Giessen

The Institutes of Human Health Protection and Animal Health Protection of the Justus Liebig University at Giessen have made an important contribution to knowledge on composting and pathogen carry-over. Giessen research is the basis for present German regulations on health hazards from the use of compost-sludge mixtures. Since public health problems will be of concern if compost is used in the United States on a major scale, it is appropriate to describe the German work in some detail.

In order to prove the positive killing of pathogenic organisms by composting, the raw refuse must be known to contain them. But finding a few residual living microorganisms in a large sample after composting is like testing for *E. coli* in treated water, a particular grab sample may or may not contain the organism. Multiple samples and most probable number statistics are therefore used in water testing. For pathogenicity tests with compost, such negatively oriented procedures are inappropriate. Rather, from an artificially implanted known number of pathogenic organisms, one must determine how many are recovered at various stages of composting. Dr. D. Strauch of the Institute of Health and Infectious Diseases of Animals has done exactly this by means of an inoculated bag technique. In this test a 100 gram sample of ground, raw refuse was put inside a bag of Perlon synthetic fiber cloth, which is not decomposed by composting. The mesh of the bag is similar to that of shirt broadcloth, preventing loss of the refuse yet allowing air to move freely through it.

Before the bag is closed, a sealed vial of pure culture of a particular pathogen is added. The pathogens in the vial are thus exposed to the same temperature but not to any microbial interaction of composting, thereby measuring the pasteurizing effects of composting. The major part of the test, however, is inoculation of the 100 grams of refuse in the bag with free pathogens.

A known number of pathogenic organisms are incorporated into this bag of refuse, either by injecting them with a hypodermic needle or by putting them into a gelatin capsule which melts at the composting temperature. (These careful procedures are necessary for safety.) These free pathogenic organisms are exposed both to the composting temperatures and to the symbiotic or antagonistic interaction of all the other microorganisms within the refuse.

In the study conducted with windrow composting, these small bags were further enclosed within a larger bag; in a study of drum composting they were protected inside a perforated steel ball to protect the bag from physical damage. Many bags were removed at various stages of composting, and the contents were cultured to determine the die-away rate of the pathogen and the duration of infection.

It was recognized that the greatest likelihood of survival would come from spore-forming pathogens, and possibly from viruses. Therefore, the pathogenic materials tested by Dr. Strauch were *Bacillus anthrax* (anthrax disease), *Salmonella enterides* (causes food poisoning), *Erysipelothrix rhusiopathise* (the swine erysipelas, a skin lymph disease), and Psittoservirus (parrot fever virus). In corollary work, typhus and typhoid pathogens have been tested. The work was done at the composting plants at Baden-Baden, Heidelberg, and Bad Kreuznach, as well as at Frankfurt on an experimental mixture of landfill residue, sewage sludge, and industrial wastes.

The results show the minimum time-temperature requirement for killing pathogenic material in compost. In windrow composting, pathogens are killed in 18 to 21 days if the temperature is continuously above 55°C (131°F). Lower temperatures require a longer time, and at Frankfurt some pathogens lived after 251 days. The amount of grinding made a difference. Kills were quicker in shredded refuse. The frequency of turning was important in that it assisted in maintaining high temperatures and better contact between pathogen and substrate. Pathogenic materials were killed in the Dano drum (using unground refuse) if held there for 5 days. Such a long detention time is not usually practical, and it was found that kill was equally effective with the normal 3 day drum detention time plus additional windrow storage for 4 days. At Heidelberg, where the refuse is ground before it is put into the Multi-Bactor tower, pathogens are killed within 24 hours.

Mineland Reclamation – Bonn

Dr. H. Kick, professor and head of the Department of Agricultural Biochemistry at the University of Bonn, has been the leader in research on the use of compost in land reclamation. Strip mining methods are used to remove lignite coal from lands west of the Rhine River near Bonn. In strip mining the overburden soil is taken off, and the lignite is then removed. The first volume of overburden is piled onto undisturbed land. Then, as the mining progresses across the countryside, the overburden of the new cut is deposited into the pit where the lignite was removed.

Typical overburden depths of German lignite mining vary from 50 ft to well over 300 ft. Since these lignite fields are dipped, they grow progressively deeper, and at some point it becomes uneconomical to strip off the overburden. At the end of the operation a substantial pit is often left, and the whole mine area is lower in elevation by the thickness of the lignite bed removed. Although German land reclamation laws require the mining company to replace topsoil to the top of the spoil area, a great deal can still be done to bring such mined lands back into production.

One particular piece of research work reported by Dr. Kick concerns a plot of land in the northwestern part of the Rhine River coal area near Cologne. The overburden soil above the soft coal was basically a silt but also contained sand and gravel. Before mining, the topsoil contained 0.55% organic matter and 35 mg nitrogen per 100 grams soil. After mining, this same topsoil did have a 1 to 1.5% carbon content, but it was recognized to be primarily waste coal. The pH was 7.3 to 7.5. This replaced topsoil had lost its structural characteristics and was susceptible to compaction and slicking over.

Plots of 25 square meters in size (270 square feet) were treated with various kinds and amounts of organic and chemical additives and fertilizers. The basic organic materials were refuse compost, dried sewage sludge, and a peatmoss sewage sludge mixture called Biohum. Chemical nitrogen, phosphorus, and potassium were added as appropriate. The plots were then planted the first year to mangels (a root crop cattle feed), then to cabbage, then to winter wheat, and finally to alfalfa for 2 years. The crop yields were all related to the check plot (no organics and no chemical fertilizers).

Yield in terms of the untreated yield ranged from 97% (where the field was overloaded by applying 125 tons of compost per acre with no balancing chemical nutrients) to 148% (on plots treated with 60 tons of sewage sludge plus chemical fertilizer). Nondefinitive appraisals of soil structure were made, and it was felt that the soils fortified with the organic materials had better physical properties.

An economic appraisal was also made; with present production and transport costs the crops were produced more economically with chemical fertilizers.

Vineyard Soil Erosion Prevention – Bad Kreuznach

Most of the vineyards that produce Germany's and Switzerland's fine quality wines are on south-facing hillsides whose slopes are often quite steep (up to 30°). The vines must be planted on the slope (not on the contour) for ease in harvesting the grapes and caring for the plants. Thus, erosion of the exposed soil between the rows of vines can be a very serious problem. For as long as farmers have been growing grapes on these hillsides, they have used organic matter to help hold the soil in place.

The usual materials have been animal manures, straw, and other crop residues. With the expansion of vineyard acreage and the increased mechanization and specialization of agriculture, it has become more difficult and more costly to obtain these usual organic materials. Compost made from municipal refuse is a good substitute and has been used in rather substantial amount. In fact, it is estimated that 60% of Germany's compost production is used in this manner, and compost probably accounts for 5 to 10% of the organic matter used in those vineyards.

Research on the use of compost in wine vineyards has been conducted at a number of places in Switzerland and Germany. Probably the most comprehensive has been at Bad Kreuznach, Germany. The present (1968) researcher is Dr. I. Bosse, a research scientist employed by the state (Rhine-Hessian); previously, the researcher was Dr. H.J. Banse.

The early research efforts were concerned with making compost satisfactorily and economically. In the early years Dr. Banse worked closely with the plant operation and with the Dano Co. in proving out the drum composting system. Later, Dr. Banse cooperated with Dr. Strauch from Giessen in the pathogen survival studies covered earlier in this report. Work begun by Dr. Banse and being continued by Dr. Bosse, relates to use of the compost. Some of this work has related to the nutrient content of compost and the nutrient availability of organically enriched soils.

An especially important and conclusive compost utilization experiment at Bad Kreuznach is concerned with soil erosion. Catch basins were installed on the 30.1° slope of vineyard soil erosion test plots. Each plot is 20 square meters (215 square feet) on the horizontal projection; this is 23.1 square meters (248 square feet) of surface exposure.

The plots are growing typical grapes farmed in the typical way except for the compost application. Compost is applied every third year at the rate of 89 metric tons per hectare (79 tons per acre); 178 metric tons per hectare (158 tons per acre); and no compost (the check plots). The surface is covered about one-half inch deep by the 79 tons of compost, and about 1 inch deep by the 158 tons of compost. The compost is mulched into the top surface a bit, but not incorporated thoroughly. Figure 2.1 shows the average yearly runoff water loss, and soil loss carried by the runoff water. The compost acts to reduce soil erosion in three ways: the crumb-like resilient compost material on the soil surface absorbs some of the energy of the falling raindrops, thereby protecting the more sensitive soil particles; the compost inherently has a high water-holding capacity, like a sponge, thereby retaining water better within the soil compost material and reducing runoff; as the compost is biologically converted to soil humus, it promotes aggregation and textural strength of the soil particles so that they resist erosion better.

This very positive benefit from using compost on hillside vineyards is the basis of compost use by the grape growers of Germany and Switzerland. Compost may or may not have secondary benefits, but reduction of soil erosion is sufficient economic justification to the grape farmers today.

FIGURE 2.1: AVERAGE WATER AND SOIL LOSS FROM 30.1° VINEYARD SLOPES

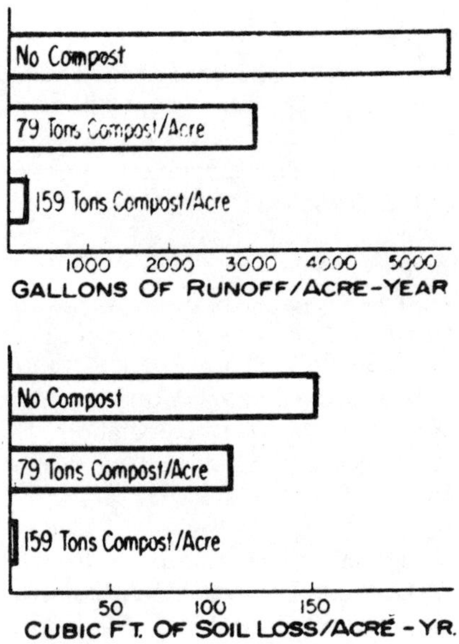

Source: PB 205,656, 1968

Compost Utilization in Basic Agriculture – Braunschweig

In all uses of compost today, the important factor is the organic matter in the compost, not the nutrient content. Before chemical fertilizers became generally available, nutrient content was important in manures, composts and other organic materials used by farmers. But, as indicated earlier, the nutrient content is very low. When chemical fertilizers first became generally available, soil research compared chemicals with organics and the latter almost invariably came out second best. Today, the research has taken on more sophistication. Research is being conducted on organics plus chemicals, the intention being to obtain maximum yield along with maintenance and improvement of the land. The Forschungsanstalt für Landwirtschaft (Research Station for Agriculture) of the Federal Republic of Germany, at Braunschweig-Volkenrode, is somewhat equivalent to the U.S. Department of Agriculture's Beltsville, Maryland Experiment Station.

One of the 12 research groups at the Research Station is the Institute for Humus Management. Research at that Institute is aimed at maintaining and improving both crop yield and soil quality through husbanding the soil organic matter. Much of the research of this Institute has dealt with compost utilization. In recent years this has been conducted by Dr. C. Tietjen.

Composts or manures added to the soil undergo biological stabilization. Byproducts of this microbial activity include humic acids and humus materials. These latter are quite effective in improving the physical characteristics of soils, through improving the soil's structural fabric. One can ask whether the applied organic matter should be well rotted or should be relatively fresh.

The argument for using well rotted materials is that one incorporates a material much like humus itself and so the soil microlife is minimally disturbed; thus crop yields are not so likely to be depressed in the year of application. The argument for applying fresh material is that the farmer supplies more microbial food, effecting a greater degree of soil improvement.

One of Dr. Tietjen's long-term compost utilization studies is concerned with this particular question. Compost directly from the Bad Kreunzach drum (5 days of rotation) is being compared with compost that has cured for 9 months after being removed from the drum. (The field plots are on heath-like sandy soil at Braunschweig; the compost is shipped from Bad Kreuznach.) Table 2.2 lists the composition of the composts used in the study.

TABLE 2.2: COMPOSITION OF COMPOST USED IN THE STUDY ON AGE OF COMPOST*

	Fresh Compost	Cured Compost
Organic content	28.4	19.1
Nitrogen	0.48	0.46
Phosphorus (as P_2O_5)	0.29	0.31
Potassium (as K_2O)	0.41	0.45
Calcium (as CaO)	2.91	3.78
Magnesium (as MgO)	0.90	1.05

*Composition given in percent.

These two composts were applied to the individual plots at 100 metric tons per hectare (89 tons per acre) at the beginning of the experiment. At the end of the first 3 year rotation, compost was applied again, at 90 metric tons per hectare (80 tons per acre), and a like 80 tons per acre was applied again between the second and third 3 year rotation.

Chemical fertilizers were applied each year: sufficient potassium to all plots; sufficient phosphorus to half of each plot and none to the other half; and nitrogen at three rates, zero, medium, and high. The effect of nitrogen is two-fold. It serves as a nutrient to the plant being grown and it fertilizes the soil microorganisms attacking the added compost. The crop rotation was potatoes, rye and oats. The soil was a silty sand with a pH of 5.8.

Table 2.3 presents the averaged yield ratios (combining potato yields of years 1, 4, and 7; rye yields of years 2, 5, and 8; and oat yields of years 3, 6, and 9). The fresh compost depressed the yield of the first crop (potatoes), whereas the cured compost effected a slight yield increase. Yields in the second and third year were higher for both kinds of compost than for equivalently fertilized but noncomposted plots, with fresh compost superior to cured compost in yields produced.

TABLE 2.3: EFFECT OF FRESH AND CURED COMPOST ON YIELDS*

	Year of application (potatoes)		2d year (rye)		3d year (oats)	
	Fresh	Cured	Fresh	Cured	Fresh	Cured
No nitrogen:						
+no P_2O_5	97	117	134	117	122	111
+adequate P_2O_5	98	117	145	128	131	123
Medium nitrogen:						
+no P_2O_5	95	104	118	111	113	111
+adequate P_2O_5	95	99	127	121	118	113
High nitrogen:						
+no P_2O_5	93	103	112	107	111	108
+adequate P_2O_5	99	100	112	108	114	107

*Yields in each treatment are expressed as percent relative to equivalently fertilized [with chemical fertilizer] but noncomposted plots

By taking the average of the averages, one can say that cured compost increased the yield relative to equivalently chemically fertilized but noncomposted plots by 11.1%, and fresh compost increased the yield by 11.7%. However, the plots well fertilized with chemicals but without compost out-produced poorly fertilized but composted plots. Thus, the farmer will generally prefer to buy and spread more chemical fertilizer rather than buy compost. The nutrient levels of the harvested crops were measured each year, and the data are averaged in

Table 2.4. In the first year, cured compost was superior to the fresh compost in increasing the nutrient content of the crop, but in the second and third years the fresh compost was definitely superior.

TABLE 2.4: EFFECT OF FRESH AND CURED COMPOST ON NUTRIENT LEVELS OF CROPS*

	Year of Application (potatoes)		Second Year (rye)		Third Year (oats)	
Nutrient in Harvested Crop	Fresh	Cured	Fresh	Cured	Fresh	Cured
Nitrogen	101	121	128	119	132	113
Phosphorus (P_2O_5)	101	103	138	120	129	119
Potassium (K_2O)	104	111	127	116	135	117

*Yields are percentages relative to noncomposted but otherwise equivalently fertilized plots.

Responses of the soil to the compost application and crop rotation were measured in the seventh and eighth years, and the data are averaged and presented in Table 2.5. As can be seen, compost additions increased the soil organic content and pH. This improved the physical structure of the soil and may well account for the better yields of composted plots than of equivalently chemically fertilized but noncomposted plots.

TABLE 2.5: SOIL CHARACTERISTICS IN THE COMPOST RIPENESS EXPERIMENT

	No Compost	Fresh Compost	Cured Compost
Organic carbon (mg/100 g)	1,950	2,300	2,280
Total nitrogen (mg/100 g)	124	153	145
Available P_2O_5 (mg/100 g)	21	31	31
Available K_2O (mg/100 g)	17	30	29
Available Mg (mg/100 g)	8	13	14
pH	5.9	6.8	6.5

In this experiment, as in others that Dr. Tietjen has conducted, it is obvious that compost does help the soil itself and does improve yields over those of equivalently fertilized (with chemicals) but noncomposted plots. The real limitation is that high applications of chemical fertilizers alone give a higher crop yield than medium levels of chemicals plus compost. Thus, the farmer is tempted to spend his available soil improvement money exclusively on chemical fertilizers rather than dividing it between compost and chemical fertilizers. Dr. Tietjen is trying to measure the long-term effects of cropping with and without composts and other organic matter.

The argument can be advanced that although chemical fertilizers may be more profitable for the short-term, soil organic maintenance through compost additions may be in order for the long-term (50, 100, 200 years into the future).

POTENTIAL FOR COMPOSTING AND COMPOST UTILIZATION IN THE UNITED STATES

The preceding sections have attempted to put composting and compost utilization into perspective. It appears that there are no real technological barriers to making compost. It does appear, however, that the utilization of compost is limited. There are successes in luxury agriculture, as evidenced in vineyard, flower, and landscaping uses, but basic agriculture cannot be expected to absorb the material. Further, the concept of composting must be considered from the municipality's viewpoint.

Most municipal officials recognize that it is not possible to make money from refuse. The most that can be expected, with regard to composting, is that the net cost to the city for refuse disposal by composting will be less, or no greater, than the costs of disposing of the garbage by landfilling or incineration, or a compensating value can be put on the factors of reduced rate of consumption of burial sites or reduced air pollution relative to incineration.

It does appear that in the United States there are three specific areas in which refuse processing and discharge through composting have potential: (1) small amounts of compost can be marketed for luxury agriculture, in which case the overall economics, to the producing municipality and to the user, are favorable; (2) the finished compost has value, but the economics, for the producer or for the user, must be related to intangible values; and (3) the finished compost is either not valuable or only marginally so, but the overall economics are nonetheless favorable.

For Luxury Agriculture

The concept of using compost in luxury agriculture has been adequately discussed in preceding sections. The success in Holland is due to an emphasis on specialty and luxury crop agriculture, and an aggressive and dynamic sales organization. The United States, with its high standard of living, also supports a significant specialty agriculture. The percentage of compost so handled would unquestionably be less than Holland's 17% but could certainly be as great as Germany's 0.5 of 1%. Possibly 1% is a realistic figure; this would amount to beneficial, economic conversion of the wastes from 2 million people.

Locations where composting operations could be successful would have to be selected with extreme care. A market analysis would have to be the first step; transport costs are so great that the compost must be marketable in the local area. Examples of this market sensitivity are pertinent. For a number of years the Metropolitan Waste Conversion Corp. at Largo, Florida, was successful in making and selling a limited amount of domestic compost.

In 1966 St. Petersburg, only 15 miles from Largo, installed a 105 ton/day refuse conversion plant which was able to make 50 to 60 tons of compost per day. This quantity completely saturated the market, so that Metropolitan Waste Conversion has ceased operations at Largo.

One might think Los Angeles would be a good area for composting because of all of the gardening and outdoor living. But refuse compost must compete there with dried and composted dairy manure, higher in nutrients and with other quality characteristics as good or better. And the manure sells for one-half to two-thirds cent/pound in bags at retail garden stores and supermarkets; compost from domestic refuse just cannot compete.

There may well be other markets, as yet untapped, for compost. With further processing, compost could be pelletized to serve as an organic carrier for chemical fertilizer. In Holland some compost is sold as a litter material for livestock housing and as a mulch for packing bulbs. There are antibiotics and vitamins in compost that might be recovered. But even with these markets it appears that the quantity of compost that could be beneficially utilized is small, the 1% suggested earlier.

Beneficial but Uneconomic

After the above emphasis on a proper economic approach, one might question the idea of compost production and utilization under a less than favorable economic situation. Compost production and utilization might be justified on the basis of lesser insult to the environment or on the basis of improvement of our already despoiled resources.

Hypothetical but specific examples of the potential of composting in this area are numerous. A city might be investigating incineration and composting. The net cost of composting might be greater than that of incineration, but climatic or geographic factors might be such that the extra air pollution from incineration would be undesirable. Composting might thus be a better choice, and the additional cost of composting would be recognized, acknowledged, and paid.

The noneconomic benefit of composting within the city's wastes management program can be cooperative rather than competitive. It is generally acknowledged that landfilling is the least expensive waste disposal method. In fact, landfilling is always a necessary part of a waste disposal program.

Yet even sanitary landfilling has disadvantages, excessive land use and possible water pollution problems. Landfilling cannot be abandoned because it may have these faults. Rather, the landfilling practice should be improved to eliminate such faults.

Composting can assist in this. Diversion of compost to other uses, of course, reduces the volume of material going into the fill. And even if refuse is made into rough compost and the compost put into the landfill (not sold), there is a volume reduction. Further, filling a landfill with uncompostable residues, or the compost itself, would be more sanitary and less likely to cause water pollution than landfilling with raw refuse. There is a second way in which composting can aid the landfilling operation. Sanitary landfills are usually used until they fill a depression or until they run out of cover dirt. Yet an above grade sanitary landfill to build a hill or mountain is very desirable for future recreational use.

One can visualize a percentage of a city's garbage being composed to produce cover material for burying the rest of the refuse. In this way a rubbish mountain, covered with trees and grass growing on compost, would benefit future inhabitants of the city.

There is a very important area where compost utilization can be beneficial although the economics are not favorable on a basis of immediate or tangible returns. This area is the reclamation of despoiled lands. Strip mining for coal, iron, gold, and other resources, has already defaced 800,000 acres of land in the United States.

It was not economic during the mining, and it is not economic today, to repair these disfigurations of the landscape. However, it is desirable to reclaim the many acres of land despoiled by strip mining, in the interest of national beautification. Organic matter is definitely needed on some of the mine spoil and debris piles, and compost from municipal waste is a logical source of this organic matter.

In the future, as recreational areas for the expanding population become more valuable, it may become feasible to reclaim naturally barren or infertile areas. Organic matter is a first need of such lands, and again, compost is an appropriate material. A specific example or a possible area of use is in the reforestation of wastelands of shallow soil that are marginally watered.

Trees will often grow on north slopes, where drying out of the soil is less severe, while south slopes are dry and barren. As shown earlier, compost improves the soil's water-holding capacity. It might well be that compost incorporation, along with forest management, could reclaim these south slopes.

All of these ideas are desirable and beneficial, but hardly economic. At least, it is hard to justify their worth through straightforward economic analysis. The intangible of future beauty does not have a generally agreed upon present dollar value. However, in the coming years some of this land reclamation and beautification will occur and will be able to utilize compost. Probably less than 1 or 2% of the nation's refuse generation would be channeled into this disposal, but it would certainly be a desirable avenue.

Nonbeneficial but Economic

There is a third and last concept under which composting and compost utilization can proceed, and this has greater potential than either of the choices discussed above. It is a middle ground, between using compost at 15 to 50 tons/acre-year for crop production or land improvement, and using the land as a refuse burial site. (About 800 tons of refuse will raise the elevation of 1 acre of landfill site by 1 foot.) Conceive of the application of 100, 200, or 300 tons of rough quality compost per acre per year, year after year. The land would be an acceptor, degrader, and stabilizer of the waste; crop production would probably not occur; but neither would the land be lost to a future use.

Each year, as much rough compost, or perhaps just ground refuse, would be incorporated into the soil as could be assimilated. When the land is needed for another purpose, the compost application would be stopped and the land would recover. This concept of compost utilization might be likened to a biological incinerator. The applied organic matter would be burned and consumed by the soil microbiological activity. The operating cost of this biological incinerator will probably be higher than running a thermal one, but the capital cost, and thus the amortized cost, should be much lower.

For many cities this scheme might well be preferred over conventional incineration. As a hypothetical example, if a soil could take 100 tons of rough compost per acre per year, produced from 200 tons of domestic refuse, this would be a land requirement of 1 acre per 500 people. This is even less land than is required for sewage lagoons.

At this stage it is impossible to estimate what percentage of the nation's refuse might be so managed. It would appear that this system might have merit for communities with populations between 10,000 and 100,000. There are 1,760 such communities in the United States, with a total population of nearly 50 million, or 25% of the nation's population. If this method of refuse disposal were appropriate for even one-third of these communities, it would be a substantial avenue of waste disposal.

Composting Practice in the United States

Material in this chapter is from PB 213,478 (1971) by A. Breidenbach and PB 215,299 (1965) by O.W. Kochtitzky and J.S. Wiley.

PROCESSES AND TYPES OF PLANTS

There are more than 30 composting systems identified by the names of their inventors or by proprietary names. In general, the systems are classified either by the method of preparation of the refuse or by the method of digestion. Sometimes both classification schemes are used in the description. In most systems, refuse is prepared for digestion by comminuting it in raspers or in various kinds of mills, including hammermills, chain mills, and wet pulpers. Sometimes a process is named for the type of mill used, such as the Buhler or the Hazemag. Digestion is accomplished in windrows, pits, trenches, cells, tanks, multistoried or multidecked towers or buildings, and in drums and bins. There are 16 types of composting processes commonly in use (Table 3.1). Present day composting plants generally provide for five basic steps in processing the refuse: preparation, digestion, curing, finishing or upgrading, and storing.

TABLE 3.1: TYPICAL COMPOSTING PROCESSES

Process Time	General Description	Location
Bangalore (Indore)	Trench in ground, 2 to 3 ft deep. Material placed in alternate layers of refuse, night soil, earth, straw, etc. No grinding. Turned by hand as often as possible. Detention time of 120 to 180 days.	Common in India
Caspari (briquetting)	Ground material is compressed into blocks and stacked for 30 to 40 days. Aeration by natural diffusion and air-flow through stacks. Curing follows initial composting. Blocks are later ground.	Schweinfurt, Germany

(continued)

Process Name	General Description	Location
Dano Biostabilizer	Rotating drum, slightly inclined from the horizontal, 9' to 12' in diameter, up to 150' long. 1 to 5 days digestion followed by windrowing. No grinding. Forced aeration into drums.	Predominately in Europe
Earp-Thomas	Silo type with 8 decks stacked vertically. Ground refuse is moved downward from deck to deck by ploughs. Air passes upward through the silo. Uses a patented inoculum. Digestion (2 to 3 days) followed by windrowing.	Heidelberg, Germany; Turgi, Switzerland; Verona and Palermo, Italy; Thessaloniki, Greece
Fairfield-Hardy	Circular tank. Vertical screws, mounted on two rotating radial arms, keep ground material agitated. Forced aeration through tank bottom and holes in screws. Detention time of 5 days.	Altoona, Pennsylvania and San Juan, Puerto Rico
Fermascreen	Hexagonal drum, three sides of which are screens. Refuse is ground. Batch loaded. Screens are sealed for initial composting. Aeration occurs when drum is rotated with screens open. Detention time of 4 days.	Epsom, England
Frazer-Eweson	Ground refuse placed in vertical bin having 4 or 5 perforated decks and special arms to force composting material through perforations. Air is forced through bin. Detention time of 4 to 5 days.	None in operation
Jersey (also known as the John Thompson system)	Structure with 6 floors, each equipped to dump ground refuse onto the next lower floor. Aeration effected by dropping from floor to floor. Detention time of 6 days.	Jersey, Channel Islands, Great Britain, and Bangkok, Thailand
Metrowaste	Open tanks, 20' wide, 10' deep, 200' to 400' long. Refuse ground. Equipped to give one or two turnings during digestion period (7 days). Air is forced through perforations in bottom of tank.	Houston, Texas, and Gainesville, Florida
Naturizer or International	Five 9' wide steel conveyor belts arranged to pass material from belt to belt. Each belt is an insulated cell. Air passes upward through digester. Detention time of 5 days.	St. Petersburg, Florida
Riker	Four-story bins with clam shell floors. Ground refuse is dropped from floor to floor. Forced air aeration. Detention time of 20 to 28 days.	None in operation

(continued)

Process Name	General Description	Location
T.A. Crane	Two cells consisting of three horizontal decks. Horizontal ribbon screws extending the length of each deck recirculate ground refuse from deck to deck. Air is introduced in bottom of cells. Composting followed by curing in a bin.	Kobe, Japan
Tollemache	Similar to the Metrowaste digesters.	Spain; Southern Rhodesia
Triga	Towers or silos called Hygienisators. In sets of 4 towers. Refuse is ground. Forced air aeration. Detention time of 4 days	Dinard, Plaisir, and Versailles, France; Moscow, U.S.S.R.; Buenos Aires, Argentina
Windrowing (Normal, aerobic process)	Open windrows, with a haystack cross-section. Refuse is ground. Aeration by turning windrows. Detention time depends upon number of turnings and other factors.	Mobile, Alabama; Boulder, Colorado; Johnson City, Tennessee; Europe; Israel; and elsewhere
van Maanen process	Unground refuse in open piles, 120 to 180 days. Turned once by grab crane for aeration.	Wijster and Mierlo, the Netherlands

Preparation

Processing of the refuse prior to composting involves several operations, which typically may include receiving, sorting, magnetic separation, grinding, and adding sewage sludge. The receiving equipment is designed to act as a refuse reservoir and to provide an even flow of refuse through the plant at the rate at which subsequent operations can process it.

As the refuse leaves the receiving area, noncompostables, bulky items and salvageable materials such as tires, large pieces of wood and metals, rags, plastics, rubber, leather, wood, glass, nonferrous metals, and paper may be removed by hand. Ferrous metals are removed then or later by magnetic separators. This sorting protects the machinery, improves the quality of the final product, and provides for salvage. Ballistic separation of heavier articles and pneumatic separation of light materials are sometimes applied after grinding.

Refuse grinding reduces particle size to facilitate handling, digestion, and mixing of the materials. Some processes, for example, the van Maanen and Dano, do not require grinding prior to digestion. In these cases, the compost is ground prior to distribution. Some recent work has been done in an effort to develop machinery capable of reducing the particle size of refuse on the composting field.

The moisture content of ground refuse is important for proper digestion. Most values given for proper moisture content range between 45 and 65% by wet weight. Work at Johnson City has indicated that 50 to 60% moisture by wet weight is needed for good decomposition. The moisture content of the ground refuse must, therefore, be adjusted to proper levels in preparation for digestion.

Raw or digested sewage sludge may be added in liquid form to provide moisture. This will also provide some additional organic, inorganic, and trace materials while providing for a sanitary disposal of the sludge. If the amount of sludge to be added is greater than that necessary as a source of moisture, the sludge must be dewatered accordingly. Other wastes, such as animal and poultry manures, and canning wastes can also be added.

Digestion

Digestion or decomposition is carried out either in open windrows or in enclosures. The principal objective is to create an environment in which microorganisms will rapidly decompose the organic portion of the refuse. Most modern plants use aerobic rather than anaerobic decomposition. In aerobic decomposition, microorganisms requiring free oxygen degrade the waste.

To furnish the oxygen, air is introduced into windrows by turning and into enclosed systems by forced draft and agitation. Heat, which is generated profusely, reaches 140° to 160°F (60° to 70°C) or higher. The heat destroys pathogenic organisms, weed seeds, fly ova, etc. Decomposition proceeds rapidly and does not produce excessively unpleasant odors.

If the decomposing mass is not aerated, the free oxygen is soon exhausted and a different microflora begins to grow. These anaerobes obtain oxygen from the various compounds in the waste and decomposition proceeds much more slowly. In the van Maanen system, the windrows are anaerobic, and the composting time required is 4 to 6 months.

By way of contrast, aerobic windrow composting takes only about 6 weeks and aerated enclosed systems only a matter of days. In anaerobic composting systems, peak temperatures are only about 100° to 130°F (38° to 55°C), foul odors arise, and pathogens may survive.

In methods having long digestion periods, the process includes a rapid decomposition stage and a ripening or curing period. In the methods involving shorter digestion, the agitation and aeration in enclosures are carried on during the earlier, more active decomposition period, and curing follows. Satisfactory stabilization is attained when the compost has the characteristics of humus, has no unpleasant odor, high temperatures are not maintained even though aerobic conditions and desirable moisture content exist, and the carbon to nitrogen ratio (C/N) is such that the humus can be applied to the soil.

Although a C/N of 20 is widely accepted as the upper limit for final application to the soil, the actual availability of the carbon and nitrogen is the determining factor and, in practice, the ratio is often higher.

The time required for digestion depends on the initial C/N if proper moisture, particle size, and aerobic conditions are maintained. Studies at the University of California on the windrow composting of mixed refuse showed the following with regard to the more active decomposition period.

Initial C/N	Approximate Days Required for Composting
20	9 - 12
30 - 50	10 - 16
78	21

If optimum conditions exist and the initial C/N is 30 to 35, refuse will take on the color and odor of humus in 2 to 5 days of active decomposition. The C/N may not, however, be lowered by the decomposition to a level satisfactory for most uses. In practice, refuse has a higher initial C/N than is considered optimum. More of the carbon is in the form of cellulose and lignin, which resist decomposition. Some authorities doubt whether materials with a high C/N or even with one above 25, can be adequately decomposed in 3 or 4 days if they contain cellulose or lignin. Thus, the high rate mechanized digesters produce a material that is given a curing period, which includes further digestion without aeration.

Curing

Curing time to permit additional stabilization depends on the use to be made of compost. If it is to be used in hotbeds, where the heat of decomposition is desirable, it can be applied as soon as the active stabilization phase is over. Compost can be applied with little curing to fields or gardens that are not to be planted for some months. If planting is to take place immediately, stabilization must have advanced to the point at which further decomposition will not rob the soil of nitrogen. In a windrowing system that calls for frequent turning for aeration, composting can be satisfactorily carried out in approximately 6 weeks with another 2 weeks for curing and drying.

Mechanical processes use various curing periods. The Dano process uses as little as 7 to 10 days storage for further stabilization after the material leaves the digester. In Aukland, New Zealand, however, where Dano digesters are also used, 3 to 4 months are given to curing. At Altoona, Pennsylvania, where a Fairfield-Hardy digester is used, the curing or maturing time is one to three weeks. The Naturizer-type plant at St. Petersburg, Florida, is reported to provide 10 days to 2 weeks for curing.

Finishing

Screening, grinding, or a combination of similar processes is done to remove plastics, glass, and other materials from the compost that might be objectionable in its use. If the compost is to be utilized as an erosion control measure in isolated places, it can be applied without being ground or screened. For the luxury gardening market, such materials must be either removed or reduced to an acceptable size. Additional upgrading, such as pelletizing or fortifying with commercial fertilizer, may also be accomplished to satisfy various markets.

Storage

The demand for compost in quantity is greatest in the spring and fall. A plant must, therefore, provide storage space for up to at least 6 months of production. The compost can be stored outdoors in piles. The storage period can, in fact, serve as the curing phase if the compost is put into low piles until heating has ceased and is then piled higher. Compost can be stored for later finishing or the finished product can be stored; it may have to be placed under cover.

SOME RECENT APPLICATIONS

European Practice

Since 1960, the literature has contained reports of about 2,600 composting plants operating outside the United States; 2,500 are small plants in India. About 100 plants have operated elsewhere, including Great Britain. Nine plants have operated in West Germany since World War II but have processed less than 1% of that nation's refuse. On the other hand, one-sixth of the refuse collected in The Netherlands is processed in composting plants. The van Maanen type plant, which was established in 1932, is still in operation and produces 160,000 tons of the country's annual total of 200,000 tons of compost.

The large number of composting plants in India is the result of an intensive program whose objective is to utilize all organic wastes on farmland. It was started by the government in 1944 and is still being supported. By 1959, the annual production of compost was 3.34 million tons. The Bangalore process is most commonly used. Hand labor, which is plentiful and inexpensive, is used extensively. Land comprises the major portion of capital costs, since the only construction needed is a series of trenches.

Operational and cost data on some European and Middle Eastern composting plants have recently been reported. Only a small fraction of municipal refuse is composted in Europe; it ranges from less than 1% in West Germany to 17% in The Netherlands.

At 12 plants studied by Kupchick, which serve a total of 3,136,000 people, 45% of the refuse processed became compost. About 70% of the product was sold. Conditions which favor sales are not uniformly distributed and result in a wide range of potential revenue. Most European cities have, therefore, selected less expensive refuse disposal methods.

Buchs and Turgi in Switzerland are of particular interest. Incineration is replacing composting there but the compost plants must remain operational so that the product remains available for those who are willing to buy it despite its high cost.

Some recent proposals and projects for composting municipal refuse in Europe and the Middle East were unsuccessful for reasons similar to those reported in the United States. The municipalities or other operating agencies did not choose to provide the additional financial support required for composting and selected a less expensive disposal alternative.

The additional support is needed to cover the increased production and utilization costs, and it might be furnished in the future if other than strict economic factors are considered. Some of the added costs might be recovered in the form of the agricultural and other benefits derived from using compost.

Some plants have such features as recycling of the compost, complicated materials-handling or processing procedures, or the use of inocula, which add to production expense; these reflect intensive promotional efforts. On occasion, provision is made to pay the municipality for the raw refuse; this is invariably an explicit warning of financial problems to come.

Information from Israel presents a mixed picture. Michaels reported that in five of the seven districts which form the State of Israel, either windrow or Dano composting plants are utilized to process refuse from 43% of the total population.

Cost figures from Europe and the Middle East are consistent with those reported by municipal compost plants elsewhere, including plants in the tropics, whether closed down as at Kingston in Jamaica or operating as at Bangkok, Thailand.

United States Practice

Prior to 1950, composting of municipal refuse received almost no attention in the United States. The need for new disposal methods, accompanied by an interest in returning organic wastes to the soil, stimulated basic studies and research on composting of organic wastes. Eighteen composting plants were funded between 1951 and December 1969 (Table 3.2).

TABLE 3.2: MUNICIPAL SOLID WASTE COMPOSTING PLANTS IN THE UNITED STATES (1969)

Location	Company	Process	Capacity (tpd*)	Type Waste	Began Operating	Status
Altoona, Pa.	Altoona FAM, Inc.	Fairfield-Hardy	45	Garbage, paper	1951	Operating
Boulder, Colo.	Harry Gorby	Windrow	100	Mixed refuse	1965	Operating intermittently
Gainesville, Fla.	Gainesville Municipal Waste Conversion Authority	Metro-waste	150	Mixed refuse, digested sludge	1968	Operating
Houston, Texas	Metropolitan Waste Conversion Corp.	Metro-waste	360	Mixed refuse, raw sludge	1966	Operating

(continued)

TABLE 3.2: (continued)

Location	Company	Process	Capacity (tpd*)	Type Waste	Began Operating	Status
Houston, Texas	United Compost Services, Inc.	Snell	300	Mixed refuse	1966	Closed (1966)
Johnson City, Tenn.	Joint USPHS-TVA	Windrow	52	Mixed refuse, raw sludge	1967	Operating
Largo, Fla.	Peninsular Organics, Inc.	Metro-waste	50	Mixed refuse, digested sludge	1963	Closed (1967)
Norman, Okla.	International Disposal Corp.	Naturizer	35	Mixed refuse	1959	Closed (1964)
Mobile, Ala.	City of Mobile	Windrow	300	Mixed refuse, digested sludge	1966	Operating intermittently
New York, N.Y.	Ecology, Inc.	Varro	150	Mixed refuse		Under construction
Phoenix, Ariz.	Arizona Biochemical Co.	Dano	300	Mixed refuse	1963	Closed (1965)
Sacramento Co., Calif.	Dano of America, Inc.	Dano	40	Mixed refuse	1956	Closed (1963)
San Fernando, Calif.	International Disposal Corp.	Naturizer	70	Mixed refuse	1963	Closed (1964)
San Juan, Puerto Rico	Fairfield Engineering Co.	Fairfield-Hardy	150	Mixed refuse	1969	Operating
Springfield, Mass.	Springfield Organic Fertilizer Co.	Frazer-Eweson	20	Garbage	1954 1961	Closed (1962)
St. Petersburg, Fla.	Westinghouse Corp.	Naturizer	105	Mixed refuse	1966	Operating intermittently
Williamston, Mich.	City of Williamston	Riker	4	Garbage, raw sludge, corn cobs	1955	Closed (1962)
Wilmington, Ohio	Good Riddance, Inc.	Windrow	20	Mixed refuse	1963	Closed (1965)

*Tons per day

As of the latter date, plants at Altoona, San Juan, Houston, and Johnson City were operating at essentially design capacity, those at Boulder, Mobile, and St. Petersburg were operating on a demand basis, and the Gainesville plant had recently closed down while alternative means of support were sought to replace the assistance previously provided under a U.S. Public Health Service grant. One plant, at New York, was under construction under a $1.3 million loan that had been provided by the U.S. Department of Commerce to provide employment in an economically stagnant area. Except for the Johnson City plant, which is a Federally supported research project, present planning requires significant sales of compost in order for the plants to be viable. The specific comments given below on several of the plants are from PB 215,299.

Altoona, Pennsylvania: For a number of years the Altoona FAM, Inc., plant composted garbage with combustible rubbish by the windrow method. In December 1963 the Fairfield Engineering Company began operation of a Fairfield-Hardy composter at the Altoona plant. This was the second installation of a Hardy digester (see Largo, Florida).

The refuse is ground in a wet pulper followed by two dewatering presses. The ground material is composted for a period of about 5 days with intermittent stirring and aeration in the Fairfield-Hardy digester. Stirring is provided by augers suspended from a rotating bridge in the circular tank. Air is provided by means of a blower and air pipes embedded in the floor of the tank.

Normal operating temperatures are about 140° to 160°F. The plant has a capacity of about 45 tpd but normally operates at about 18 to 23 tpd. Successful experiments have been run in which digested sewage sludge was added to the ground refuse. The compost is reported to sell readily as a soil conditioner or as an organic base for fertilizer. However, the company buying compost from the Altoona plant ships most of it in bulk a distance of more than 500 miles.

Largo, Florida: The original Largo plant was built by Peninsular Organics, Inc., and operated for a few years using a Hardy composter. In 1963, National Composters Company refinanced Peninsular Organics Company and rebuilt the Largo plant using a digester of their own design. The plant has a capacity of 50 tpd. Operating 5 days a week it treates 40 to 50 tpd of mixed refuse to which is added 1,000 to 1,500 gpd of essentially raw sewage sludge.

The steps involve sorting and magnetic separation, double hammermilling, mixing of ground refuse and sludge, composting for 6 days in a long narrow open tank with turning and aeration, regrinding, screening, and bagging. The digester is 12 ft wide by 6 ft deep by 280 ft long having a metal false bottom with air holes. The turner is a rail mounted bucket elevator moving in either direction the length of the digester. Material is turned twice during composting. A test for pathogens by the Florida State Board of Health Laboratory gave results which were quite favorable.

Norman, Oklahoma: The first plant using the Naturizer process was built a number of years ago by the Naturizer Company of Norman, Oklahoma. This plant and the later one at San Fernando, California, were taken over subsequently by SACS, Inc. (Salvage and Conversion Systems), which entered into a marketing agreement with Westinghouse Electric Corporation.

In February 1965, these companies, together with Allstate Insurance Company, formed a new company, International Disposal Corporation, which incorporated the assets of SACS. The plant treated the 35 tpd of mixed refuse from the city of Norman. Operations were discontinued in 1964 except for experimental purposes. Two unique features of the plant were a specially-designed swing hammermill and the Naturizer composter. The digester system provided 6 day retention in two three-story buildings with a movable floor in each story. Compost was sold either in bags or bulk. Considerable time and effort were expended to build up a market for the product, with sales activity extending as far away as Dallas.

Phoenix, Arizona: The Phoenix composting plant was the second Dano-type plant in the U.S. and was built by Arizona Biochemical Company to process 300 tpd of refuse. Originally two Dano biostabilizers were provided but two additional units were added in March 1963. Even then, by European practice of 3 day digestion in the biostabilizers, the capacity of these rotating drum units is only about 175 tpd. The plant provides sorting, magnetic separation, salvage, hammermill grinding, composting in the biostabilizers for 24 to 30 hours, screening, regrinding, and deglassing, and further composting in windrows.

Sacramento County, California: The plant in Sacramento County, built by Dano of America, Inc., in 1957, was the first Dano-type plant in the U.S. It had a capacity of about 40 tpd of refuse and, except that only one biostabilizer unit was employed, operated in much the same way as the Phoenix plant. The plant was closed in 1963.

San Fernando, California: The San Fernando compost plant was built by Lockheed Aircraft Corporation and operated by General Conversions System Corp. Later, the plant was taken over by SACS, Inc., and finally by International Disposal Corporation, Shawnee, Oklahoma (see above). The plant had a capacity of 70 tpd of refuse but operated at only about 40 tpd.

It was an improved version of the process developed at the Norman, Oklahoma, plant. Appreciable salvage operations were performed. It was necessary to provide an afterburner on top of the composting unit to comply with Los Angeles County Air Pollution Control District requirements to ensure that no odors would be discharged.

The plant was shut down in October 1964. Compost was sold in bags and in bulk. A unique development for the use of compost was called Sta-Soil which was a water suspension of compost, chemical fertilizer, grass seed, and shrub seed. It was sprayed over denuded slopes, or cuts and fills, to provide a blanket against soil erosion and a soil in which grass and shrubs would quickly take root.

Springfield, Massachusetts: The Springfield plant was built in 1954, partially burned down in 1958, and was rebuilt in 1961. It was a Frazer-Eweson process plant owned by the Springfield Organic Fertilizer Company. The plant had a capacity of 20 tpd of garbage from the city of Springfield. The garbage, being too wet to compost aerobically, was held for about two days to dewater. This frequently created fly, rat, and odor problems. The city failed to renew the contract in 1962 and the plant was closed.

Williamston, Michigan: The Riker System plant at Williamston was built in 1955 and operation was taken over by the city in 1956. The 4 tpd plant treated garbage, vacuum-filtered raw sewage sludge, and corn cobs. Ground garbage and corn cobs, mixed with the sludge cake, were composted for 21 days in two 4 compartment vertical composters. The plant was closed in 1962 primarily because a nursery in Detroit stopped purchasing the compost.

Wilmington, Ohio: The 20 tpd refuse composting plant at Wilmington was built by Good Riddance, Inc. A minimum of sorting is provided and two hammermill grinders are used in series. The ground refuse is composted in windrows where it is turned weekly by means of a front end loader for an indefinite period of time. Difficulties have been experienced in selling the finished compost.

ENGINEERING, CHEMICAL AND MICROBIOLOGICAL ASPECTS

As systematized and mechanized composting operations were developed, engineering problems increased. Various digestion arrangements were developed and patented, and some work was done on special grinders. In most cases, the material-handling equipment used had been developed for other industries and modified to process refuse.

Although considerable laboratory or small-scale work has been done in the last 20 years in the United States and the basic technologies are known, adequate experience in design and operating compost plants has not yet been accumulated. As a result, most plants have gone through a period of cutting and trying with different types of machinery and plant layouts before going into production.

The laboratory work done on the physical and chemical aspects of composting serves as a basis for process control in full-scale plants. The extreme heterogeneity of raw refuse and other factors, however, result in the composting of mixed municipal refuse being practiced, in some respects, as an art with laboratory research serving as a guide. This section discusses the general engineering, chemical, and microbiological aspects of composting, based on observations made and data accumulated over nearly two years at Johnson City and a year at Gainesville.

Engineering Aspects

General Design Criteria: A number of criteria must be considered in designing and operating a compost plant. One is to obtain all the information possible about the population to be served and the amount and type of refuse it generates. For example, a domestic refuse high in cellulose may make the material resistant to attack by microorganisms, and the composting process may have to be changed accordingly.

On a national scale, seven pounds of urban (domestic, commercial, institutional, and municipal) solid wastes are generated per capita per day. This figure includes garbage, rubbish, trash, ashes, demolition debris, street sweepings, dead animals, abandoned vehicles, etc.; it does not include industrial or agricultural solid wastes. The amounts collected vary according to seasonal, climatic, and socioeconomic factors. Production rates for individual areas must, therefore, be determined by surveys. A second design criterion is the length of the workweek. Thus, a plant operating on a 5 day workweek is required to accept refuse at 1.4 times the rate for a 7 day design capacity.

Another factor is the number of shifts to be worked per day. To process equal amounts of material, a plant operating on two shifts does not need some of the large refuse-handling machinery or grinders that a one shift operation has to use. The receiving area must, however, allow for storage for processing during the second shift of about one-half of the refuse delivered to the plant during the day. Digestion, storage, and curing elements must be sized for the total tonnage received.

Refuse Handling: Plants must provide an area appropriately designed for receiving refuse and large enough to store at least one day's delivery. The refuse moves from the receiving area to size-reducing equipment, frequently via a picking station, where salvageable items, noncompostables, and large items that might damage equipment are removed.

The flow of refuse from the receiving area should be controlled. Some hoppers are discharged to an oscillating belt to achieve this control while others may use a leveling gate. Arching or bridging often occurs in the receiving hopper and may be more acute if a leveling gate is used. The operation often proceeds more smoothly if one or both of the hopper's long sides are nearly vertical. If the incoming refuse has been compacted, as in a transfer trailer, it must be broken up and pushed into the hopper. A front end loader has been successfully used for this purpose.

Endless moving belts are widely used to carry refuse from station to station. When hand picking is practiced, the bed of refuse should not be more than 6 inches deep; belt width and speed are the determining factors. If the belt is too wide, the pickers cannot reach its center. If the belt traverses any space outside a building, covers must be provided. They must be easily removable and high enough and wide enough that refuse does not catch on them. Sideboards or skirts should be used to keep refuse from falling from the belt.

Ground refuse moves more easily than raw. The belts should be wide enough or have sideboards to prevent spillage and minimize cleanup problems. Bucket elevators work well in lifting ground refuse, and screw feeds can be used to move it horizontally in troughs. Narrow openings, restrictions, or chutes must be avoided because ground refuse clogs easily.

In freezing weather, it may be necessary to heat the belts where they come in contact with the end pulleys. Wipers should be installed on the belts near the drop-off points, so that refuse, especially ground refuse, does not stick to the returning undersides and drop on the floor. Hoppers and bins that hold refuse or ground refuse only temporarily should have moving belts in their floor or have openings large enough for refuse to be gravity-pulled through the bottom.

Separation of Noncompostables and Salvage: Most plants remove as many noncompostables (wood, plastics, glass, metals, rags, etc.) as possible before the refuse reaches the size-reducing equipment. If this is not done, some picking of bulky items is necessary, either at the receiving point or from a belt, to protect the equipment.

When salvaging is practiced, the material removed is usually classified, and an effort is often made to remove paper. At Johnson City, where no salvaging is practiced, two pickers can handle up to 60 tons of refuse in 6 to 8 hours. In Gainesville, where paper and metals are salvaged, six pickers are used to process 125 tpd.

At most plants, ferrous metals are removed by magnetic separators. These may be in the form of a permanently magnetized head pulley installed on the raw or ground refuse belt or an overband type that uses an electromagnet. If two grinders are used in series, the magnetic separator may be located between them.

Rejected material at Johnson City has averaged 26% by weight of the incoming refuse. At Gainesville, about 10% is removed as salvaged paper while another 10 to 30% is rejected. Some composting plants are trying to salvage up to half the incoming refuse by using special mechanical devices. Rejected, unsalvageable material must be moved to a disposal site. A market is usually available for paper and metal, and cans, glass, and certain plastics can be sold in some areas. In Europe, refuse often has a high ash content. Rotary and vibrating screens are sometimes used to remove the ash from raw refuse before it is ground.

Comminution: Refuse is usually ground or shredded to improve materials-handling and digestion operations. Most of the machines now used were originally designed for use with homogeneous types of materials. The most common grinding device is the hammermill. It usually consists of high-speed swing hammers connected symetrically on a horizontal shaft and cutter bars that have grate openings through which the refuse is forced.

Refuse fed into the mill is comminuted by the application of high tensile and shearing forces. Tensile force is applied as the swinging hammers flail the refuse against the breaker plates. The shearing forces come into play as the hammers force the refuse through the grate openings. Hammers are of various types, and some are better suited than others to produce the shredding action needed. Several types of double rotor mills have been developed in Europe, and at least one is manufactured in the United States.

Hammermills require relatively large motors and must have the capacity and power to handle a flow of refuse that resists grinding; the capacity depends on the particle size desired. It is common practice to use two mills in series; the first produces a rough grind while the second reduces the particles to 2 inches in the largest dimension. Refuse is abrasive and the hammers must be frequently refaced. It has been found at Johnson City that the hammers need rebuilding after 30 to 40 hours of use. Since hammermills operate at 1,200 to 3,500 rpm, they produce noise and vibration. The machines should, therefore, be mounted on dampening materials and the feed chute should be flexible or have a flexible connection. A specialized shredder or rasper developed in The Netherlands consists of a large vertical cylinder that surrounds a vertical shaft on which heavy arms are mounted. They rotate horizontally above a perforated floor. Pins or studs, mounted in panels on the floor and along the sides of the cylinder, shred the refuse, and the particles then fall through the perforations. The revolving arms are hinged and swing when they meet resistance.

Raspers operate more slowly than hammermills and require less power, but they have a greater initial cost and require more floor space. Performance data indicate that the perforated plates and pin plates in the 10 tons/hr (rated capacity) rasper at Johnson City need replacing after grinding about 10,000 tons of refuse (approximately 1,500 operating hours).

Refuse must build up in the grinding compartment for about 20 minutes before effective grinding begins. If the flow of refuse stops, the machine runs at a diminishing rate of production until empty. It should, therefore, be kept full throughout the day for the greatest efficiency. The perforated floor acts as a sieve and retains oversize material that can be discharged at intervals through a chute. Raspers must be cleaned out frequently, but they are so designed that workmen can easily enter the grinding compartment.

Since large pieces of dry cardboard may build up in the machine and overload it, water is sometimes sprayed on the refuse either before it reaches the rasper or after entering it. This procedure may prove disadvantageous if sewage sludge is to be added after grinding, because the refuse may become excessively moist if the sludge is not sufficiently dewatered.

Wet pulpers, such as the one at Altoona, Pennsylvania, where cans, bottles, and other noncompostable items are not normally received in the garbage are also used to comminute refuse. They consist of a large bowl that holds a rotatable steel plate studded with hardened steel teeth. After the bowl has been partially filled with water and the plate is rotating at about 650 rpm, raw refuse is dumped in. It is whirled against the teeth and shredded. The resulting slurry, which contains about 5% refuse solids, is subsequently discharged through a horizontal bar screen. It must be dewatered by 40 to 50% to be digested.

Addition of Sewage Sludge and Other Organic Wastes: Sewage sludge may be satisfactorily composted along with a community's refuse. The cost is about the same, in some cases less, as for conventional systems that use anaerobic digestion, drying beds, and subsequent disposal. It is usually mixed into ground refuse in mixing drums.

When using sludge, the water content of the ground refuse sludge mixture will normally be greater than that desired for composting unless the sludge is dewatered somewhat. It is not practical to use sludge prior to rasper operations because it contaminates the refuse, which may have to be later cleaned from the rasper. Water is, therefore, often added before and during the grinding process. The amount used has an effect on the sludge dewatering operation. When a hammermill is used, water is added after grinding, and all of it may normally be obtained from sewage sludge. Raw sludge is preferred to digested sludge because it can be dewatered more readily and has a higher nutrient content. (Digested sludge can, nevertheless, be used.) The amount of dewatering necessary depends on the ratio of sludge to refuse to be processed and the initial water content of the sludge and the refuse as received.

Depending on the amount to be removed, dewatering can be accomplished in gravity tanks equipped with vacuum filters, in centrifuges, or by using rotating cell gravity filters. Gravity tanks with picket agitators may suffice in many cases. In humid climates, water is removed mechanically from sludge and refuse. Adding other organic wastes to municipal refuse before it is composted appears feasible as a method to dispose of such wastes. The composting process is apparently not affected, and the nutrient contents of the compost may, in fact, increase.

Efforts have been made to have water added automatically by an electrical-mechanical system, but such techniques have not worked well. Experienced plant operators can often tell by the mixture's appearance and handling characteristics when the optimum moisture range of 50 to 60% has been reached.

Digestion: Aerobic composting or digestion is carried on in windrows or in such enclosures as aerated tanks or bins. The success of any aerobic method depends on aeration, mixing, and maintaining the proper moisture content. In most plants, efforts are made to maintain aerobic conditions to avoid odors, obtain higher temperatures, and achieve more rapid decomposition.

Experience has shown that unground refuse can be composted, but normally it is first ground so that the particles average 1½ to 2½ inches in their largest dimension. This encourages rapid decomposition either in windrows or in enclosed systems. At Wijster and Mierlo in The Netherlands, however, unground refuse is windrowed according to the van Maanen process, which calls for only one turning; composting takes 4 to 6 months. In the Dano system, the refuse usually introduced into the digester is unground. The constant turning of the drum reduces the size of the particles as they are digested. Where windrow turners are used, they may also shred the material as they mix it.

In the windrowing process, aeration and mixing can be accomplished by using a front end loader or a clamshell bucket on a crane. Turning machines with a shoveling or screw arrangement are also used. These turners are designed to pick up the material from a belt and place it on the ground. Another type turning machine, with a rotating drum on which teeth are mounted, straddles the windrow and turns it in place.

Some preliminary turning experiments conducted at Johnson City indicated that the windrow should be turned at least once a week. Two turnings per week produced the best decomposition; more frequent turnings proved less efficient because temperatures in the windrows dropped. (Higher temperatures are needed to destroy pathogens.) The degree of decomposition obtained was determined on the basis of appearance, odor, and low carbon content. In windrow composting where supplemental aeration is not normally provided, the moisture in the material must be kept at 50 to 60% by wet weight to keep maximum decomposition proceeding. If the moisture content is higher, water fills the voids in the compost and slows the biological process by denying it sufficient oxygen. On the other hand, dry windrows may cool and fail to decompose properly; water is, therefore, incorporated into the mass.

In wet weather, the windrows may have to be turned frequently to help release the moisture. Too much wetness may cause the decomposition to become anaerobic and give rise to odors. At Johnson City, windrows normally remain in the field for at least 6 weeks and temperatures of up to 160°F are maintained. The compost is then moved to a curing shed where it is allowed to dry for 2 weeks or longer. Experience has indicated that high relative humidity will prevent satisfactory air drying.

In enclosed composting systems, forced or natural draft air is provided for digestion. The material is intermittently turned in the tank by a special apparatus or constantly turned by mixers, rakes, or the rotating digester. Digestion takes 3 to 10 days; the longer period produces a more stable product.

As in windrow composting, insufficient oxygen in an enclosed digester creates odors and slow digestion. Water content must be maintained at between 50 to 60%. This level may be higher if means for efficient air transfer have been provided. Temperature profiles are comparable to those observed in windrow composting.

At Gainesville, the refuse is kept for about 2 weeks in two parallel digestion tanks, each 330 ft long, 20 ft wide, and 10 ft deep. Air is periodically introduced through perforated plates in the bottom. The tanks are equipped with movable conveyors for removing the compost; the conveyors can also mix the material but are not used for this purpose.

Curing: The period of active, rapid, digestion is followed by a slower stabilization period, called curing. In the windrowing process, if proper conditions for decomposition are maintained, digestion and curing form a continuum. Compost is usually removed from the field and cured under cover. It is then ready for many uses, but further stabilization or curing goes on for months.

Finishing: Compost can be used for various purposes as received from the windrowing field or digester. Often, however, it does not have uniform size particles and may contain bits of plastic, glass, or other nondecomposable objects. It is usual practice, therefore, to finish the compost by regrinding and screening it. When these steps are taken, the moisture content should not exceed approximately 30% by net weight.

This may vary, however, depending on the finishing process used and the desired results. In proper climatic conditions, air drying alone may yield a product dry enough for satisfactory finishing, but mechanical dryers may have to be used in humid and wet areas. At Johnson City, air drying has proved difficult all year, especially during wet winter months. Hammermills may be used for regrinding. Screens can be rotary or vibrating types and have perforated plate, square mesh, or piano wire type screening elements with openings up to ½ inch. In the last type, the transverse wires (which are very taut and are perpendicular to the flow of compost) can be at least ¼ inch apart and the longitudinal supporting wires up to 10 inches apart.

Regrinding can precede or follow screening. In the latter case, the material retained by the screen is sent to the grinder and then screened again. Small particles of glass, whose presence is usually objectionable, can be removed by machine using one or more of the principles described later.

For some uses, such as land reclamation or erosion control in isolated places, compost need not be finished. For general agriculture, a coarse grind is satisfactory, whereas for horticultural and luxury gardening the product must be finer. Reground and screened compost is ready for use as a soil conditioner or may serve as a carrier for fertilizers and blended products. Pelletizing, especially with blending, is sometimes done.

Storage: The use of compost in quantity is seasonal, being more in demand during the spring and fall. A plant must, therefore, be able to store its production for 6 months or more. Curing and storage can be combined by piling the compost after its heat has diminished or disappeared. Rough compost can be stored for later grinding or the finished product may be stored. Storing in the open may be feasible in some cases.

Special Problems

Glass Removal: Glass removal presents a problem. Pieces and articles of glass are broken as the refuse is collected and transported to the plant as well as by the receiving and processing machines; complete removal is, therefore, impossible. Glass crushers, often simply two spring-loaded rollers that exert pressure on each other, are sometimes used to break the material into small sizes. Hammermills can pulverize glass particles to some extent, but a rasper's capability is minimal.

Many European plants have an apparatus that uses gravity and the differences in the inertial energy and resiliency of particles to remove glass. A ballistic separator impels the material horizontally or at a slight upward angle. Dense and resilient particles travel farther than those that are soft and nonresilient. Although the separation is not definitive, it is satisfactory. The secator relies on gravity and particle elasticity to remove heavy and resilient bits of material.

The bounce plate is so positioned that the compost or ground refuse lands forward of the center of rotation of the drum and is carried to the far bin. The resilient particles bounce off the plate to hit the drum back of the center of rotation and bounce into the near bin. In the inclined conveyor separator, the belt is made of steel plates. Heavy and resilient particles bounce down while softer ones continue upward and are deposited in another container.

Another type of separator, known as a stoner, employs a diagonally inclined, perforated, vibrating table or plate. The material to be separated is deposited on the plate and is fluidized by an upward flow of air through the plate. The lighter particles are thereby separated from the heavier ones and are transferred across the plate, then down to a discharge point. Heavier particles are carried upward and discharged at the top.

Plastics Removal: Removing plastic film and similar items may also present special problems. Some film can be removed by pneumatic devices, but their development has not been perfected. Dense plastic particles also give trouble. Small, flexible items can be deformed to allow them to pass through a hammermill or a rasper, after which they resume their shape in the ground refuse. Salvaging molded plastics is being investigated in some areas.

Handling Problem: Compost requires special material-handling techniques. It tends to stick to chutes, sides of hoppers, inside surfaces of dump trucks, etc. One operator in this country has used a Teflon compound on the inside surfaces of dump trucks that carry large quantities of compost. Bulk shipments in railroad cars present unloading problems because the compost will not flow by gravity from conventional cars, as do coal or crushed stone.

Weight and Volume Losses: 20 to 30% (by net weight) of the incoming refuse is not compostable, and some of this is removed. The remaining refuse is comminuted to aid the digestion process, which, in turn, further reduces the volume. The weight lost is in the form of the two principal products of decomposition, carbon dioxide and water; it amounts to 20 to 30% of the dry weight. Experience gained at Johnson City and Gainesville indicates that each ton of incoming refuse will yield, after processing, about 1,000 lb of compost having a moisture content of approximately 30%.

The volume reduction achieved in composting has created considerable interest in preparing refuse for landfilling by grinding it or by grinding and then composting it. In addition to occupying less space, the ground material has other apparent advantages: it looks better than raw refuse, does not contain large pieces of paper that can blow about, and is less attractive to rodents.

If it has been composted as well as ground, the refuse has an even better appearance, gives off fewer odors, restricts fly breeding, requires less or possibly no cover, and occupies less area. Since it has been digested, the compost, if well composted, should subside less and produce less gas than raw refuse.

It has been estimated that if a given amount of raw refuse were divided into equal parts, one of which was buried untreated in a landfill and the other was first composted, the latter would occupy 21% less space. Another source states that if refuse containing noncompostables is ground and then composted, it can double the life expectancy of the standard sanitary landfill for a given depth of fill. Work being done at Madison, Wisconsin has shown that milled refuse, compacted to a depth of 6 ft with a D-8 bulldozer, takes up only about half as much volume in a landfill as unmilled refuse handled in accordance with usual sanitary landfill practices. Further reduction in volume may be achieved by using special compactors. It is likely that if the material had also been composted, even less space would have been required. At Johnson City, 42 day old compost has 28% less volume than ground but uncomposted refuse. This compost, however, does not contain the proportion of noncompostables contained in the previously mentioned raw refuse.

These observations indicate that if refuse is milled (except items that could jam or damage the machinery) and then composted, its volume is reduced by at least half. Composting costs in this case would be reduced as there would be little sorting, compost could be removed from the digesters as soon as a practical point of decomposition had been reached, no curing or drying period would be needed, and no finishing would be required.

Environmental Aspects

Composting plants may affect the surrounding environment and the neighborhoods in which they are situated, because they are potential sources of odors and may provide breeding places for flies and rodents. Good management, especially the maintenance of aerobic conditions in the composting refuse, can, however, minimize the odor problem. Managers should insist on meticulous housekeeping and avoid holding unground refuse from one day to another.

Adult flies and fly larvae and pupae are brought into a plant with the refuse, especially if the collection system does not provide frequent pickups. At the receiving point, the application of a residual insecticide around the unloading apron and on the walls of the receiving building has successfully killed larvae migrating from the refuse. Grinding also destroys many of the larvae and pupae.

Flies are also attracted to fresh ground refuse, and they may breed during the digestion period if proper conditions are not maintained. On the other hand, the temperatures reached in aerobic composting are lethal to fly larvae and eggs. Care should, therefore, be taken to ensure that all portions of the windrows reach these temperatures.

This can be done by proper shaping and piling prior to turning. If the windrows are turned approximately every three days, this may also aid in controlling flies by breaking their life cycle. The judicious use of an insecticide will also help. Rodents can be controlled with poisons and by denying them hiding places.

Noise and dust may be hazardous to the workers. Since hammermills can generate intolerable noises, they should be isolated from the building by dampening materials. Materials falling into a metal-sided reject hopper from a picking station may also cause excessive noise. Lining with wood or some other soft material can ameliorate this condition. In areas where much coal is burned, ash impregnated refuse may be a problem because of the dust generated. The same could be true if street sweepings are part of the refuse.

Chemical Aspects

Carbon-Nitrogen Relationship: The rate at which organic matter decomposes is determined principally by the relative amounts of carbon and nitrogen present. In living organisms, the ratio is about 30 to 1 and, theoretically, this should be the optimum ratio in municipal refuse also. In actual practice, however, it is much higher. Composting, nevertheless, can successfully create a product suitable for agricultural use, since it is pathogen- and nuisance-free and is produced in a reasonable length of time from refuse having initial carbon to nitrogen ratios ranging from 21 to 78.

As composting proceeds, the causative organisms use the carbon for energy and the nitrogen for cell building. The C/N becomes smaller with time, since the nitrogen remains in the system while the carbon is released as carbon dioxide. If fresh or insufficiently decomposed compost, with high carbon and low nitrogen values, is applied to soil, the continuing microbial activity could, in theory, rob the soil of nitrogen if the ratio exceeds 20:1. In practice, however, a higher ratio can be tolerated if the carbon is not readily available to the organisms, i.e., is in the form of paper.

Experience at Johnson City indicates that refuse with an initial ratio of between 39 and 49 will decompose in about 6 weeks into a compost with a ratio of between 28 and 35, a median reduction of 27%. The product is safe with respect to health, has a satisfactory appearance and odor, and is comparable to that produced by other plants and systems. In preliminary experiments at Gainsville on refuse and refuse-sludge mixtures, the initial ratios generally ranged from 57 to 68. After digestion, the span was 54 to 59, a 6 to 14% reduction.

Composition: The composition of compost varies widely, and data have been collected on the values of certain constituents observed at Johnson City (Table 3.3). Carbon, nitrogen, phosphorus, potassium, sodium, and calcium occur mostly in a combined form; iron and aluminum, and possibly magnesium and copper, are present primarily as uncombined metals. The values found for nitrogen, phosphorus, potassium, calcium, and percent ash correspond to those found by investigators of other composts.

Gotaas has reported that the organic content of compost is between 25 and 50% by dry weight; at Johnson City, it has been 60% and above for finished compost. Compost is not a fertilizer but is comparable to a good topsoil because of its nitrogen, phosphorus, and potassium content. Since it has a high organic content, it helps to provide good tilth, water-holding capacity, and nutrient-retaining capacity when mixed with poor soils.

Although such elements as iron and aluminum occur in relatively high amounts, they are present as metals and metal oxides and should not pose any problems. Aluminum is a major constituent of most soils, and causes difficulties only in very acid soils, those with a pH well below 5.0. As is the case with fertilizers, liming agents, and other materials placed on the soil, consideration should be given to the effects of soluble salts present in compost and drainage must be provided so that they do not accumulate in the soil.

TABLE 3.3: ELEMENTS IN 42 DAY OLD COMPOST AT JOHNSON CITY

Element	Containing Sludge (3 to 5%)	Without Sludge	Range (all samples)
	--- Percent Dry Weight (average)---		
Carbon	33.07	32.89	26.23 - 37.53
Nitrogen	0.94	0.91	0.85 - 1.07
Potassium	0.28	0.33	0.25 - 0.40
Sodium	0.42	0.41	0.36 - 0.51
Calcium	1.41	1.91	0.75 - 3.11
Phosphorus	0.28	0.22	0.20 - 0.34
Magnesium	1.56	1.92	0.83 - 2.52
Iron	1.07	1.10	0.55 - 1.68
Aluminum	1.19	1.15	0.32 - 2.67
Copper	<0.05	<0.03	
Manganese	<0.05	<0.05	
Nickel	<0.01	<0.01	
Zinc	<0.005	<0.005	
Boron	<0.0005	<0.0005	
Mercury	not detected	not detected	
Lead	not detected	not detected	

Moisture: To achieve the greatest decomposition, the water content of compost should be maintained at 50 to 60% by wet weight, and aeration should be provided. As water is added, the compost becomes more compact and this reduces the amount of air present. Anaerobic conditions then arise and objectionable odors are created. If too much water is introduced, the material becomes difficult to handle and too dry for finishing. On the other hand, if the moisture content falls below 50%, high temperatures are achieved in the center of the mass and it gives off few odors, but the rate of decomposition slows.

Temperatures: Temperature readings made in a composting mass may indicate the amount of biochemical activity taking place. A drop in temperature could mean that the material needs to be aerated or moistened or that decomposition is in a late stage.

It has been noted that the windrowing method produces a typical temperature profile. Temperatures between 150° and 160°F (66° to 71°C) are easily reached and maintained for about 10 days. Temperatures between 140° and 150°F (60° to 66°C) can be kept for about 3 weeks. Temperatures of up to 170°F (77°C) have been observed in the center of a composting mass.

Time-temperature relations are important in freeing the compost of pathogens. At Johnson City, it has been found that a single weekly temperature reading will help determine if composting is progressing normally and that temperatures necessary to destroy pathogens are being maintained.

At Gainesville, the compost has sometimes reached 180°F (82°C) on the sixth day of composting in open-tank digesters. Forced aeration is used at this plant, but agitation is provided only intermittently or not at all.

On the Fairfield-Hardy digester at Altoona, Pennsylvania, temperatures between 140° and 160°F (60° and 71°C) are normally attained and occasionally rise to 176°F (79°C). In this enclosed system, the composting material is continually agitated for 7 to 9 days; forced aeration is used.

pH: The initial pH of refuse at Johnson City is usually between 5 and 7 unless a large amount of alkaline material is present. On an average, the refuse is at least three days old when it arrives. The pH drops to 5 or below in the first two to three days of composting and then begins to rise; it usually levels off at about 8.5 and remains there as long as aerobic conditions are maintained.

If the compost becomes anaerobic, as it does when stored in deep piles at Gainesville, the pH drops to about 4.5. Ordinarily, pH is not used for process control, but if an operator knows the normal pattern it follows, he may be alerted to the presence of unusual substances if differences are noted.

Microbiological Aspects

General: Composting as a microbiological process is the conversion of biodegradable organic matter to a stable humus by indigenous flora, including bacteria, fungi, and actinomycetes, which are widely distributed in nature. In composting, however, such selective factors as moisture content, oxygen availability, pH, temperature, and the carbon/nitrogen ratio determine the prevalence and succession of microbial populations.

As Waksman, Gordon, and Hulpoi have pointed out in extensive studies on the aerobic composting of manure and other organic matter, a variety of microorganisms has a number of specific functions, all of which are interrelated in the total process. During the course of composting, both qualitative and quantitative changes occur in the active microflora; some species multiply rapidly at first, change the environment, and then disappear to allow other populations to succeed them.

When composting begins, the mesophilic flora [microorganisms able to grow in the 77° to 113°F (25° to 45°C) temperature range] predominate and are responsible for most of the metabolic activity that occurs. This increases the temperature of the composting materials, and the mesophilic populations are replaced by thermophilic species, those that thrive at temperatures about 113°F (45°C). This rise in temperature is influenced to a great extent by oxygen availability. When municipal refuse is composted at Johnson City, for example, windrows kept for the most part aerobic reach temperatures up to 167°F (75°C) and produce few objectionable odors. When a windrow is allowed to become anaerobic through lack of turning, however, the temperature peaks at about 130°F (55°C) and drops much lower after the first two weeks of composting.

Even though composting materials usually contain a wide range of active flora, many attempts have been made to develop an inoculum of microorganisms that would speed the decomposition process. Their use has, however, usually proved to be of little value. Nevertheless, it would seem worthwhile to study the merit of adding nitrogen, phosphorus, or other elements to supply essential nutrients for the active flora in the composting of straw, paper, and other materials that, alone, are nutritionally unbalanced.

The key to successful composting in the United States may well depend on acquiring the ability to degrade the increasingly high concentrations of cellulose found in solid wastes. Advances in this area appear to depend on the gathering of more knowledge about the functions of specific flora in the composting process, a field in which relatively little research has been done.

Pathogen Survival in Composting: Studies conducted at Johnson City and by Morgan and MacDonald indicate that properly managed windrow composting turns out a product that is safe for agricultural and gardening use. Proper management consists of keeping the moisture content at between 50 and 60% by wet weight, maintaining aerobic conditions by turning the material periodically, and assuring that the windrows are thoroughly mixed. Specifically, investigations made at Johnson City in conjunction with East Tennessee State University showed that:

> Pathogenic bacteria that may be associated with sewage sludge and municipal refuse were destroyed by the composting process after being inserted into windrows;
>
> There was a consistent, inverse relationship between the number of total and fecal coliforms in the compost and the windrow temperatures recorded. A heat range of 120° to 130°F was sufficient to reduce the coliform populations significantly, often to a level at which they could not be detected by the Most Probable Numbers Method. Significant numbers of coliforms reappeared, however, when the temperature dropped during the last stages of the composting process;
>
> *M. tuberculosis* was normally destroyed by the 14th day of composting if the temperature had averaged 149°F (65°C). In all cases, the organisms were destroyed by the 21st day;
>
> Composting that attains a temperature range of 130°F or higher for as little as 30 minutes also deactivates the polio virus;
>
> There are no references in the literature to any sanitation workers having been infected by fungi as a result of handling solid wastes. This suggests that there should be no restrictions put on the use of compost.

No extensive studies regarding pathogen survival in mechanical composting systems in the United States have been completed, but there are indications that the product is safe to use if it has been properly mixed in a mechanical digester-composter and then cured.

ECONOMIC CONSIDERATIONS

Composting in the United States has not been looked upon as a method of waste disposal but as a business; as such, it has had an unsuccessful history. Considering it as a disposal method, the expectation of a profit or an income to balance the cost adds a burden not imposed upon landfilling or incineration.

Thus, one deterrent to more widespread development of composting as a means of municipal solid waste treatment in this country has been this widely advanced premise that composting must produce a profit, or at least pay its own way. No other method of waste disposal or treatment is expected to accomplish such a goal.

In the last 20 years, the technology of composting municipal refuse has been investigated rather intensively, and there is the knowledge and equipment to enable engineers to design mechanized compost plants and to produce compost. Although corresponding information on costs is much less satisfactory, it has become increasingly apparent that composting is not an inexpensive method of refuse treatment.

This section considers the monetary aspects of composting. Further research in the use of compost in agriculture and land management may help to furnish a gauge by which to measure economic benefits not now quantified. Elements of the cost of disposal by composting, expressed as a gross cost per ton for processing raw refuse, and the credits that may accrue from salvage, the sale of compost, and other considerations are discussed.

A reader attempting to discover the cost of composting is confronted with an array of costs ranging from about $2.50 to $20.00 per ton of refuse processed. Cost figures for individual plants are available but variations in size, methods of operation, plant complement and wage scales, number of shifts, accounting systems, financing details, land costs, and final disposal make comparisons almost impossible. Until recently, the principal source of such information was Europe. To apply costs developed in Europe or elsewhere to composting in the United States is even more difficult.

Because of this lack of reliable cost data on the construction and true operating costs of composting plants in general, the major portion of the information that follows is based on observations of the U.S. Public Health Service, TVA Composting Project, Johnson City, Tennessee, although it is limited to the general conclusions and aspects of costs as derived from these observations.

Capital Cost

Windrowing Plants: Estimates of the capital costs for various capacity windrow composting plants, based on the actual costs encountered for the Johnson City composting plant, range from $16,560 per ton of daily capacity for a 50 tpd plant to $5,460 per ton of daily capacity for a 200 tpd plant on a two-shift operation. The estimates of the total yearly capital investments for these plants on the basis of cost/ton of refuse processed, range from $6.15 for the 50 tpd plant to $2.01 for the 200 tpd plant on two shifts.

The actual initial capital costs for the USPHS-TVA Composting Plant were $18,580 per ton daily capacity. On a per ton refuse processed basis, the yearly capital investment cost is $12.98 (at 34 tpd in 1968). Operated at the design capacity of 52 tpd, the yearly capital investment cost would have been $6.88 per ton refuse processed.

The capital cost of $965,980 for the Johnson City plant is subject to some qualifications. A high proportion (38% of plant cost) is in buildings, partly because of the multistory design with equipment installed on the second and third floor levels. More ground level floor space and simpler framing, as used in common mill buildings, with installation of machinery independently of the structure would have permitted a less expensive structure. Similar reductions were used in the cost projections for the other plants. A case in point is the 150 tpd plant at Gainesville, Florida where the cost of the building, estimated at $150,000, is approximately 11% of the total plant investment.

These cost estimates include equipment for processing sewage sludge from the population generating the refuse. Since these composting plants include sludge processing equipment, caution must be exercised; costs developed here cannot be directly compared with capital costs of landfills or incinerators that do not include equipment for sludge processing.

Enclosed Digestion Plants: Enclosed digestion plants are similar to windrowing plants with respect to receiving, sorting, grinding, adding sewage sludge, final grinding and screening, curing and storage. Plants of both types require area for the storage of compost for curing and stockpiling. Seasonal use of compost makes stockpiling necessary. The estimates for the windrow plants include land for storage, in rectangular piles 15 ft high, of 6 months' production. Land required for the composting area is also important. Land costs used in the estimates were $800 per acre; this figure is consistent with land values near the Johnson City plant. By way of comparison, land near the Gainesville plant costs about $4,000 per acre.

Comparing the capital costs per ton refuse processed for the digestion systems of a 150 tpd windrowing plant with those of an enclosed type plant, shows that, although the windrowing plant requires more land, capital cost per ton processed will be less for a reasonable range of land prices. Many of the other costs associated with these plants would be similar.

Other Countries: Capital costs reported in 1965 for European plants vary from $0.76 to $1.91 per ton of raw refuse processed using the windrow methods. For enclosed systems, the range was $1.18 to $3.98.

It must be noted again that it is difficult to compare plant costs because of such factors as variations in size, type, and operation. Comparisons with foreign plants are even more difficult. The complexity of construction will, of course, influence costs. In warm climates, heating of buildings may not be necessary. For windrowing plants, the size and spacing of the windrows will influence land requirements. In wet periods and in humid climates, mechanical dryers may have to be installed.

Operating Costs

Windrowing Plants: Estimates of the yearly per-ton-of-refuse-processed operating cost for windrow plants of varying capacities, again made by projecting the actual costs encountered in operating the composting plant in Johnson City, ranged from $13.65 for the 50 tpd plant to $8.70 for the 200 tpd plant on a two shift operation.

Actual costs for operating the Johnson City composting plant in 1968 were $18.45 per ton of refuse processed. The nature of the research conducted there and the inability of the Johnson City municipality to deliver enough refuse for operation at full plant capacity are some of the reasons for the seemingly high cost.

A cost of $13.40 per ton of refuse processed was projected for operating this plant at full design capacity (52 tpd) in 1969, with some modifications for the research work being conducted. Labor expenses for 1968 amounted to about 75% of the operating costs. In 1969, they accounted for approximately 78%.

Up to 30% of the refuse delivered to a compost plant is noncompostable. If salvaging is not practiced, all of this material should be disposed of in a sanitary landfill. An estimated cost of from $0.50 to $1.00 per ton of refuse processed must then be added to operational costs. If the compost must be eventually disposed of in a landfill, the additional cost per ton of refuse processed may reach $0.50.

High-Rate Digestion Plants: Operating cost data for many of the high-rate digestion plants is incomplete, adding to the difficulty in comparing costs. Yearly operating costs per ton of refuse processed for the Gainesville plant were $7.56 for 157 tpd and $6.94 for 346 tpd. Operating costs for some European plants have ranged from about $1.51 to $2.76 per ton of refuse processed.

Total Cost of Composting: The estimated total costs per ton of refuse processed for various composting plants ranged from $3.85 to $20.65. The range for windrowing plants, estimated from data obtained from the USPHS-TVA project, however, was from $11.23 for a 200 tpd plant to $20.65 for the 50 tpd plant. The total cost for the high-rate digestion plant at Gainesville was estimated at $10.53

per ton of refuse processed at 157 tpd and $8.58 per ton of refuse processed at 346 tpd. The $32.31 per ton cost of composting municipal refuse at the USPHS-TVA composting plant is subject to the qualifications as stated in the discussion of its capital and operating costs. The projected cost of $21.16 per ton of refuse processed at full operating capacity is also subject to the same general qualifications.

Partial Recovery of Costs

The cost of composting municipal refuse may be reduced in several ways. Direct returns are possible if compost and salvageable material are sold. An indirect benefit may derive from processing sewage sludge with the refuse and disposing of it as a component of the compost.

Compost Sales: The price at which compost can be sold depends on the benefits to be obtained from its use and what customers are willing to pay for such benefits, which have yet to be accurately ascertained. Benefits from using compost over a number of years and residual benefits over a period of time from one application may increase its value.

Compost has been sold for horticultural use, and viniculture may offer a market in some areas. Conditioning or improving the product by screening, pelletizing, bagging, and providing well-planned sales promotion and distribution may result in a greater gross return. Compost may also be sold in bulk, finished or unfinished, as well as fortified with chemical fertilizers.

The University of California estimated in 1953 that farmers would pay from $10 to $15 per ton; in fact, they showed little interest. A plant in San Fernando, California, sold compost in 1964 at $10 per ton. Other sources estimated a bulk selling price of $6.00 per ton in 1967. In 1968, a St. Petersburg plant attempted to sell compost for commercial agriculture at $9.00 per ton.

The Lone Star Organics Company, Houston, Texas, was reported by one source to have sold compost at $12.00 per ton and at $6.00 per ton by another source. The Gainesville plant has sold compost for about $7.00 per ton. This was for a ground, unfortified, unpelletized product. Altoona FAM, Altoona, Pennsylvania, sold a pelletized product in 1966 and 1967 for $16.50 per ton (bulk basis) and $42.50 per ton in 40 lb bags. In the 1967 and 1968 season, orders were taken at $20.50 per ton in bulk.

Sale of Salvaged Materials: The income from salvaging depends on the cost of salvaging operations, the volume of salable materials, and the prices paid for the recovered materials. There may, however, be no market for salvaged materials in some localities. One source has stated that salvage can be practiced to at least the break-even point if a 200 tpd capacity plant is located near an industrial city. Materials most easily salvaged for which a market often exists are paper, metals, rags, and glass. There may develop a market for some type of plastics. Actual data on the income possible from salvaging are few. The plant at Gainesville is equipped to salvage and market paper and metals.

In 1968, paper was sold at $15 to $20 per ton. Shredded cans at destination could have been sold for $20 a ton, but shipping charges made this impractical. Although few rags were salvaged, they brought $18 per ton at the plant, baled. Projections for this plant have shown an expected net income from salvaged paper of $1.50 per ton of refuse processed. A feasibility study for a 300 ton capacity composting plant in Michigan assumed that paper would be salvaged in the amount of 15%; metal and cans, 9%; and glass, 10% of incoming refuse.

Composting Sewage Sludge with Refuse: A composting plant may be operated to obviate part of the cost of handling the sewage sludge received from the population it serves. For a 200 tpd plant processing all of the sludge from the population generating the refuse, the estimated savings could range from nothing to $35 per ton of sludge solids, depending on degree of treatment. Based on this estimate, the credit to composting would be from nothing to $1 per ton of refuse processed. These estimates are based on data from the windrowing plant at Johnson City. Savings might be greater for plants using high-rate enclosed digesting systems.

Composting and Landfill Operations: There is interest in reducing landfill requirements by grinding and composting refuse prior to depositing into the fill. The crushing of cans and bottles, the reduction in size of other noncompostables, and the reduction of the volume of organic material by digestion will reduce the volume of the refuse.

The digestion results in a less noxious material, less gas production in the fill, and possibly less subsidence. The compost is less attractive to rodents and insects and its appearance is more acceptable to most people. Less cover will be needed as it may be applied only to prevent a fire hazard and to keep small pieces of plastic film, shards of glass, and bits of metal from showing, as compost for landfilling would not be finished to remove these.

It has been stated that with good compaction, the landfill volume required will be about half that required for well compacted, unground refuse. More work will be required in this area on the compactibility of compost. Organic materials tend to be springy on compaction.

Landfill sites are becoming scarcer near urban centers especially due to the resistance of citizenry to such operations. When sites are found at greater distances the same difficulty is often experienced where people do not want the city's refuse disposed of in their area. Composting may offer a solution in some cases. The reduction in volume can result in savings in handling costs, and sites nearer to cities may be tolerated where predigested material is deposited.

It would thus appear that composting may effect savings where hauls are long, but will not provide savings in land costs unless they are very high. Where the availability of land is the problem and not the cost, composting could extend the life of landfills. Well digested, but unfinished compost could be used for fills in many places in a community and the unused product could be put into landfills.

The recovery of salvageable materials would reduce the volume of material to be composted and to be disposed of by landfill.

Net Cost

The net costs estimated for the windrowing plants range from $18.65 per ton of refuse processed for a 50 tpd plant to about $7.73 per ton of refuse processed for a 200 tpd plant on two shifts. Net costs for the high-rate plant at Gainesville, Florida, were estimated at $6.90 per ton of refuse processed at 157 tpd to $3.45 per ton of refuse processed at 346 tpd.

Composting Costs Compared to Sanitary Landfill and Incineration

Even with an income from compost and, in some cases, from salvage sales, most composting plants show a deficit or an expected deficit. Based entirely on economic considerations, most composting plants would not be able to compete with sanitary landfilling as a refuse treatment method.

As with compost plants, the operating costs reported for incinerators vary greatly, due to the same factors that cause differences in composting costs, land values, labor costs, residual disposal, etc. For incinerators constructed after 1950, averaging a daily input of 375 tons, the operating costs have been reported at $3.27 to $4.05 per ton. A 168 ton/day incinerator reports a total cost of $6.40 per ton processed, while a 300 ton plant is estimated at $5.20 per ton.

Although references to incinerators with costs between $3,000 and $5,000 per ton of daily capacity can be found, those now being planned are more complicated and costly because of new or contemplated air pollution control measures. Costs to achieve these new criteria may have the effect of almost doubling the price for small incinerators and adding at least 30% to the cost of larger plants. An 800 tpd plant considered for Washington, D.C., was estimated at $4,500 to $5,400 per ton of daily capacity. The additional cost per ton for installing air pollution control equipment was $2,800 to $3,700.

In comparing costs of compost plants to incinerators, note that a direct comparison is not correct for incinerators operating continuously for 24 hours, as most of the compost plants considered operate only on one 8 hour shift. Also, the composting plant cost includes sewage sludge processing equipment not included in incinerators.

Thus, although the capital costs for composting plants are greater than those for landfilling, they fall in the range expected for incinerators. Some compost plants in the 300 tpd size range may equal some incinerator costs without the benefit of income from salvage and compost sales. Indications are that many will not. The 150 and 200 tpd plants may compete economically with incineration if there is an assured market for compost and salvaged materials. Plants under 100 tpd capacity appear uneconomical. The accurate prediction of a market for compost and salvage materials and the intensive cultivation of this market is thus essential in determining the economic potential for a given compost plant and will help determine whether incineration is less expensive than composting for a given community.

AGRICULTURAL AND HORTICULTURAL UTILIZATION

Agricultural Productivity and Soil Erosion Control

An excellent review of plant and soil relationships and the results of studies on compost utilization are contained in a recent paper by C. Tietjen and S.S. Hart, "Compost for Agricultural Land?", *Journal of the Sanitary Engineering Division,* Proceedings of the American Society of Civil Engineers, 95 (SA2):269-287, April 1969. The following discussion of benefits and limitations of composting related to agricultural productivity and soil erosion control draws heavily upon that paper.

Plants can grow in almost any type of soil, but soil fertility is closely related to the amount of organic matter it contains and particularly to the amount of nitrogen present. Organic matter includes humus, living plant roots, bacteria, fungi, earthworms, insects, etc. When a virgin soil is cultivated without being fertilized, its organic content and yield are reduced with time.

High productivity can be maintained if manures or chemical fertilizers are applied in the amount and at the time the crop needs such nutrients. Over long periods, higher yields result from the use of combined chemical and manure fertilizations. This was confirmed over a nine year period in which chemical fertilizers with compost added were applied to soils. Increased crop yields may, however, be obtained more economically if chemicals alone are added.

With regard to supplying plant nutrients, compost neither performs as well as chemical fertilizers nor meets the legal requirements established by several states for designation as a fertilizer. A typical compost contains approximately 1% nitrogen, ¼% phosphorus, and ¼% potassium. The slightly higher values that result when sewage sludge and municipal refuse are composted together are derived from the sludge.

The type of soil is an important factor to be considered in evaluating how the continued use of a chemical fertilizer will affect productivity. If the soil is low in organic matter, the continued use of chemical fertilizers that do not have an organic amendment may decrease crop yields over a period of time. The benefits of using compost to supply organic matter to various types of soils, and the other benefits that might be derived from its continued use over a long period of time have not been adequately defined.

Tietjen and Hart point out that yields are not the only consideration in evaluating the benefit of compost. They report the following additional information on the 9 year experiment mentioned above. The nutrient levels of the crops were measured each year. Potatoes grown on composted plots averaged 6%

more nitrogen, phosphorus, and potassium per pound of crop harvested than those grown on uncomposted but fertilized plots. On an average, compost grown rye and oats had 4 and 9% higher nutrient contents, respectively. These are significant increases.

Organic matter affects the physical characteristics of soil. Benefits that may be obtained by the addition of humus (from compost) to soil are improved workability, better structure with related resistance to compaction and erosion, and increased water-holding capacity. Improved workability is generally described as tilth; it is measured by the farmer in terms of easier plowing or cultivation which results in savings of power and time.

Better structure and improved water-holding capacity are particularly important for erosion control on steep slopes. Comprehensive research on erosion control of hillside vineyards was conducted by Banse at Bad Kreuznach, Germany. The results of his field tests on compost applied every three years to a 30° vineyard slope showed that compost was very effective in reducing erosion.

Tietjen and Hart indicated that it is difficult to put an economic value on compost applications for improvement of soil physical properties. They concluded that an improved water-holding capacity has not yet been related definitively to either increased yield or reduced irrigation requirement, nor has improved soil workability been related to a lower plowing and cultivation cost.

In basic agriculture, maintenance of acceptable soil physical properties and prevention of erosion are obtained economically through such practices as crop rotation (often with legumes, green manuring, contour farming, and fallowing). Although compost application might improve soil physical characteristics or erosion control still further, an economic analysis to prove the worth of composting has not yet been made.

The preceding examples of potential benefits from compost utilization are derived from Europe where compost has been used more extensively than in the United States. There is, therefore, a need for quantitative data on its costs and the benefits in this country.

Although there has been considerable speculation about the values of trace elements, qualitative evidence indicates that the benefits derived result from the humus component when compost is applied to lawns. There is insufficient information regarding commercial agriculture.

Demonstration and Utilization

Demonstration: Demonstration projects of the joint USPHA-TVA composting project are described in Chapter 7. Three other soil improvement demonstrations of that project deserve special mention. Two involve erosion control and the reclamation of strip mine spoil bank areas. One project is being conducted in cooperation with TVA's Strip Mine Reclamation Section and the other with the Southern Soil Conservation Committee in Mercer County, West Virginia.

In the third demonstration, approximately 100 tons of compost were shipped to Oak Ridge National Laboratory and used as a soil amendment to help establish a growth of white clover for special ecological studies. Radioactive solid wastes had been buried at the site under very poor soil, and earlier efforts to grow vegetation on this soil had been unsuccessful.

During the first 16 months of operations at the Gainesville plant (March 1968 to June 1969), 17,514 tons of compost were produced and 1,774 tons were sold. Another 5,841 tons were donated for various public uses, leaving over 55% to be stockpiled or disposed of in some manner. The proximity of the St. Petersburg compost plant has undoubtedly restricted the amounts that can be utilized, and some compost was shipped up to 170 miles away.

It has been applied at rates varying from 1 to 10 tons per acre at citrus groves, 16 tons per acre for strawberry crops and up to 100 tons per acre for pine and fern seedlings. Observations indicate that growth, crop yield, and erosion control improved. Long-term information is required to determine benefit-cost relationships. Some results from Northern Florida, however, have indicated that at least 20 tons per acre of compost must be used to achieve meaningful benefits.

Horticultural Utilization: The demonstrated benefit of compost applied to lawns has been previously mentioned. The luxury market, which includes private lawns, gardens, golf courses, hothouses, and similar applications, is governed by an entirely different set of factors from those that apply to agricultural markets. The luxury market is small-scale, labor-intensive, more sensitive to aesthetic, conservationist, and emotional considerations, and less able to evaluate extravagant promises of benefits that are claimed by advertisers of competing products. In contrast, large-scale agriculture is characterized by the need for showing profits over short periods of time comparable to that considered by other industries.

POTENTIAL

With present technologies of solid waste production and disposal, together with currently effective economic and environmental constraints, most communities are not willing to fund the cost of composting their municipal refuse. Other sections have identified the factors upon which this decision is based.

The Problem: In 1967 there were an estimated 260 million tons of solid wastes generated by urban domestic, commercial, institutional, and municipal sources. A 1970 estimated level of approximately 300 million tons, with a 50% yield, would provide 150 million tons of cured compost. (The other 50% would be accounted for almost equally by weight lost during composting and material sorted from the incoming refuse as salvage or rejects to be disposed of separately.)

Cured compost typically contains 30% water and weighs 600 pounds per cubic yard. The volume of the 150 million tons of compost produced would, therefore, be 500 million cubic yards. The fraction of municipal compost that can be marketed depends upon the costs of producing and applying it, relative to the benefits derived from using it. Compost is not a fertilizer but a soil conditioner. Some feel that its important value lies in its organic matter, which may improve the physical properties of the soil.

It is generally accepted that the cost of composting and the need to enrich the product or supplement it with chemical fertilizers restrict its marketability to buyers in the specialty fertilizer field. In this respect, municipal compost is in competition with aged cattle manure from dairies and feed lots and with peat moss.

Agricultural Effects from Compost Utilization: Although there are some benefits and some drawbacks associated with the utilization of municipal refuse compost, the economic realities associated with commercial agriculture or horticulture, which would be affected the most, have discouraged the widespread production and consumption of compost. Even barnyard manures, which are relatively rich in nitrogen, have become a disposal problem because their assumed cost-benefit ratios compare unfavorably with those of chemical fertilizers.

Organic materials, including compost, have been cited by Kilmer as the nearest thing to a cure-all for soil problems that we have. Municipal compost, however, is at a disadvantage, because it has low nitrogen values and contains plastic and glass fragments.

A dilemma results from accepting the validity of compost systems, they turn out a product that may have some values but they cost more to operate than the end product is apparently worth. McGauhey suggests that this dilemma be solved by postponement. Conversion of a low-value waste material that nobody wants into a low-value resource that nobody wants should be deferred. This can be done, McGauhey suggests, by placing solid wastes in landfills until their value warrants mining and recovering them.

Bowerman has recommended that composting be applied to regional solid waste management in the Fresno, California area. He proposes that poultry and livestock manures with low carbon-nitrogen ratios be mixed with municipal refuse and composted. The product, along with that resulting from fruit and vegetable processing wastes, would be applied to the land at a rate of 75 tons per acre per year. According to Bowerman, 20% of the nation's municipal refuse could be processed and disposed of in this way by the year 2000.

Digested sewage sludge could also be disposed of onto the land. Except for its suggestion that a market might be developed for the compost, the proposal is an example of a rather advanced systems approach to regional solid waste disposal problems based on existing technology.

In Resource Systems Management: Resource systems management is defined as directing and maintaining the development and utilization of air, water, mineral, and living resources and their interactions under steady state conditions. This means that proper incentives and recycling technologies must be found to ensure that elements, compounds, mixtures, and total energy maintain essentially their historical distribution in time and space. The economics of scale that are utilized in resource development, processing, transportation, and disposal become diseconomies at that point at which materials are finally returned to the environment. These diseconomies are minimized by returning residuals to the environment through dispersed rather than concentrated mechanisms.

Engineering control can provide greater initial dilution or dispersion. For example, a large number of factory chimneys or stacks discharging steam and carbon dioxide to the atmosphere is preferred on both economic and environmental grounds to a single stack through which a combined discharge of carbon dioxide and water would go. Modern sewers that discharge sewage treatment plant effluents or cooling waters into marine or lake waters have multiple discharge ports spaced over perhaps a half mile, not just a single port at the discharge end.

On land, farming of digested sewage sludge, oily sludge from refinery operations, or livestock manures promotes more rapid assimilation by the environment than if these wastes are concentrated in a small area. The organic residual of municipal refuse may also be rapidly assimilated by the soil provided that it is dispersed and has good physical, chemical, and sanitary characteristics. Compost is amenable to such initial dispersion and assimilation.

The factors that will influence the future of the composting process as a municipal solid waste management tool are the costs and benefits of the process, as compared with other municipal solid waste management processes. The present and potential technology of composting will permit organic materials to be recycled back into the soil without significantly polluting water or land. The cost is, however, higher than that associated with other acceptable management methods. On the other hand, changes in designated priorities on the use of land, sea, or air may occur as per capita waste generation rates rise.

The potential usefulness of all solid waste management systems, including those that employ composting, will be influenced by changes during future decades of the value assigned to, or the emphasis placed upon, any of the following four factors: the acceptance of more stringent standards for environmental quality; the availability of systems to meet these standards; cost per ton of solid waste managed for each available system; public policy decisions requiring beneficial recycling rather than land or sea disposal of wastes.

Composting Dewatered Sewage Sludge

Treatment and final disposal of sewage sludge represent a major cost in building sewage treatment plant facilities. Two major methods presently used for treatment and disposal of sewage sludges are anaerobic digestion and incineration. Other methods such as burial of dewatered sludge and application of undewatered sludge to farmland have severe limitations. When the problems involved with these presently used disposal methods are considered, a reduction of volume and weight would be very beneficial. Composting dewatered sewage sludge offers this possibility.

Composting of refuse has had many problems. Some of these problems are associated with the costs of separating and grinding large volumes of refuse and obtaining a market for the final product to offset some of the costs of disposal. The addition of sewage sludge to refuse composting is said to accelerate the decomposition and improve the final compost. It is commonly accepted that dewatered sludge (containing 70% moisture) from a community represents only about 15 percent by weight of the total solids disposal problem. The work described in this section illustrates the ability of composting to effectively treat dewatered sewage sludge alone to produce a stable, hygienic material and is derived from Public Health Service Publication No. 1936, 1969 written by G.L. Shell and J.L. Boyd of The Eimco Corporation.

Composting of sewage sludge alone is a new approach to a very old problem. Preliminary work indicated that weight, volume, and solids reduction by this treatment method appeared very attractive. A test program was designed to obtain the following basic information concerning the composting of a combination of primary and secondary dewatered sewage sludges: (1) effect of mixing; (2) effect of moisture content; (3) effect of recycling; (4) process capacity; (5) air requirements; (6) effect of chemicals used for conditioning the sludge for dewatering; (7) composition of off-gases; (8) destruction of pathogenic bacteria, fungi, nematodes, and viruses; and (9) chemical and physical composition of the final compost.

All of the 16-month study, except that related to the destruction of pathogenic organisms, was performed by The Eimco Corporation Sanitary Engineering Research and Development staff under Director A.A. Kalinske. The work on the destruction of pathogenic organisms was performed at the University of Utah, School of Medicine, Department of Microbiology, under the direction of Dr. Bill Wiley.

PROCESS DESCRIPTION

Composting is an aerobic thermophilic biological process. As in any aerobic biological process, an adequate method for providing oxygen and thorough mixing are required to obtain efficient odor-free treatment. Composting of sewage sludge, as studied here, necessitated three phases of operation: dewatering the sewage sludge, mechanical composting, and final curing. Each of these phases is described with reference to a flow diagram (Figure 4.1), and the dewatered sewage sludge was fed to the composter as indicated in the feeding schedule described below.

Dewatering – The sludge used during this study was a combination of primary and two-stage trickling-filter secondary sludge from the Salt Lake City Sewage Treatment Plant. The sludge was obtained each day and stored in a 350-gallon storage tank. Before vacuum filtration, the sludge was conditioned with ferric chloride and lime or a polymer to aid filtration. The sludge was then dewatered on a vacuum filter and the cake was ready for feeding to the composter.

Feeding Schedule – Dewatered sewage-sludge filter cake was fed into the composter 8 to 10 hours per day, 6 days per week. This schedule was chosen to approximate the feeding schedule that would probably be encountered in most small-size sewage treatment plants. Most smaller plants filter during only part of each day and accumulate sludge the remaining time. Six days of operating data were averaged to present results on a weekly basis. No feeding or data collection was made on Sundays.

Mechanical Composting – The dewatered sludge was then mixed with recycled compost from the composter to adjust the moisture content and to thoroughly seed the incoming sludge with aerobic thermophilic microorganisms. This mixing was accomplished in a screw conveyor used to feed the composter. The combination of sludge filter cake and recycled compost then entered the mechanical composter where it was continually mixed (by paddle mixers) and aerated for about 20 days or more depending on the final volume of compost removed from the composter. As the result of biological action during the composting period, the temperature of the mixed bed averaged about 140°F.

Final Curing – Each day the final product from the composter was stored in an individual pile, and it was then observed for 2 weeks.

FIGURE 4.1: COMPOSTING DEWATERED SEWAGE SLUDGE

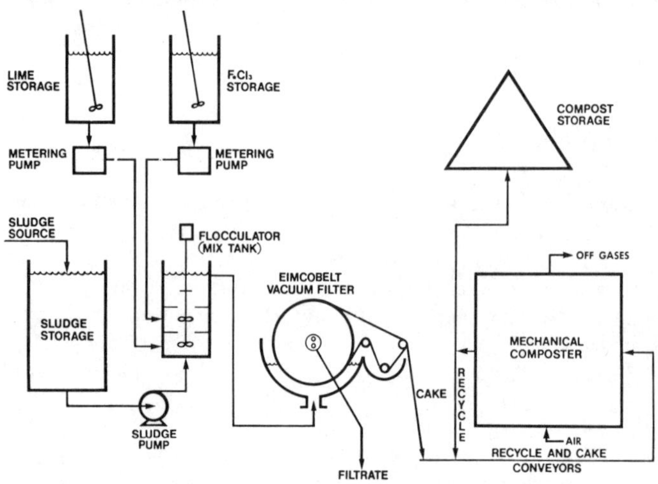

PILOT PLANT DESCRIPTION

The pilot plant was located at the Salt Lake City Sewage Treatment Plant inside the digester heat-exchanger building. A pipeline was connected to the plant's main sludge line, which went from the primary clarifiers to the digesters. Sludge was drawn from this line by line pressure to fill a 350-gallon storage tank. This storage capacity represents about 500 lb of filter cake having 70% moisture.

The sludge was then continuously pumped with a variable speed Moyno pump from the storage tank to a flocculator. The pumping rate was varied according to the capacity of the vacuum filter to dewater the sludge.

The flocculator consisted of a 20-gallon baffled tank having a variable speed, 6-inch diameter propeller. Chemical conditioners (lime and ferric chloride or polymer) were fed into the flocculator by metering pumps. The ferric chloride and lime were added to the sludge to give a content of 4.5 and 15.5%, respectively (based on sludge dry solids). Polymer was used in later tests and was added to the sludge to give a content of 1.6% on the same basis. The conditioned sludge then overflowed into the vacuum filter where it was dewatered.

The vacuum filter consisted of a vacuum pump, filtrate receivers and pumps, and the filter itself. The filter drum was 3 ft in diameter and 1 ft wide, with an effective filtration area of 9.4 sq ft. It was operated at a 30% drum submergence and at an average drum speed of 8 mpr (minutes per revolution). During cake formation, the vacuum was maintained at 12 to 15 inches of mercury and during drying, at 22 to 23 inches of mercury. The filter cake was discharged into a screw conveyor that moved the cake to a collection drum. At 30 minute intervals, a portion of the collected filter cake was weighed and dumped into a second screw conveyor. At the same time, a portion of composted sludge was also weighed, recycled, and mixed with the filter cake with the use of the same screw conveyor.

The mechanical composter consisted of an open top vessel having outside dimensions of 4 by 4 by 4 ft (40 cu ft effective volume), a porous carborundum bottom, and four 20-inch diameter paddle mixers rotating at 9 rpm. The inside of the composter was baffled and shaped to minimize cake packing to the sides and corners and to prevent short circuiting. Steel fingers were attached to the inside walls to make the mixers self-cleaning. A double lobe-type blower supplied air to the composter; an orifice-manometer system employing water in the manometer measured the air; and a mercury manometer measured line pressure.

The operating level of the composter was maintained at a selected reference level. Each morning, before feeding, compost was removed, weighed, and placed in a storage pile for observation. The amount removed each day was determined by reducing the compost bed level to the reference level. If the initial bed level was below the reference level, no compost was removed. The effluent compost was stored in an outdoor area, and each area was marked and dated for further observations.

SAMPLING AND TESTING

Sampling and data collection for tests were conducted twice daily except for compost temperature and power readings. Compost temperature data were taken 4 to 10 times daily, and power data were measured once each week. Samples of compost were taken at the outflow of the composter, whereas filter cake samples were taken directly from the filter discharge. Both types were discrete (grab) samples. The application of each test technique to the tested materials was checked with known standards for feasibility and accuracy.

MIXING

Because of the high organic content of the dewatered sludge (requiring much oxygen) and its tendency to pack, continuous mixing was necessary. The major problem was to find a mixer so designed that it would not plug but would tend to lift and loosen the material. The ribbon augers that were tried initially clogged rapidly. Several modifications of these augers were tried without success. Finally, paddles, slightly angled so that lifting would occur, were attached to the drive shaft. Mixing appeared satisfactory, and later testing proved it to be good.

Because the compost had a tendency to pack, dead areas and clearances between the mixer and side walls and bottom had to be minimized. This was done by shaping the internal part of the composter as closely as possible to that of the mixers.

Wearing of the mixer paddles indicated that the abrasive qualities of the compost material were significant. A great deal of wear occurred on the mild steel paddles during a 3-month run. The environment in the compost bed was also conducive to corrosion; moisture, heat, and oxygen promoted corrosion and aggravated the wearing problem.

In composting garbage and refuse, mixing and aeration are used intermittently. In this study, both mixing and aeration were continuous. Tests were made to determine if continuous mixing was required. Observations made when aeration alone was employed showed that the compost bed temperature began to drop within one-half hour after mixing was stopped. It was believed that continuous mixing was required to keep the bed loose and to prevent channeling of air.

Several tests were made to determine the extent of mixing that occurred in the composter. In two separate tests, plastic cubes and cylinders (approximately 250) were placed in the feed to the composter. These materials were dispersed throughout the compost bed within 2 hours. This rapid dispersion was not due entirely to agitator mixing, but was also related to the amount of compost recycled. The higher the recycle volume, the more completely mixed was the compost bed. As described later, a high recycle volume was very beneficial to the treatment process.

MOISTURE CONTENT

Moisture content of the filter cake and the compost was determined daily during this study. The moisture content of the dewatered sewage sludge depended upon the discharge characteristics of the vacuum filter, the raw sewage sludge characteristics, and the chemical conditioners used. The moisture content of the compost depended upon the filter cake moisture, retention time within the composter, and the rate of the aeration and mixing.

When ferric chloride and lime were added for conditioning at values of 4.5 and 15.5%, respectively (based on sludge dry solids), the filter cake had a moisture content of 68 to 74% and the material was easily handled and conveyed. Filter cake conditioned with Rohm and Haas C-7 to give a content of 1.6% (based on sludge dry solids) had an average moisture content of 77%. A moisture content of 70% can be realized with polymers, but this would require a more complex application technique.

Moisture content in the composter ranged from 20 to 35% with an average of 26%. The moisture content, as stated before, depended on many variables, all of which may be controlled to some degree. The amount of filter cake and its moisture content determined the quantity of moisture entering the composter daily. Retention time and rate of aeration affected the degree of evaporative drying. Moisture content of the feed to the composter depended on filter-cake and recycled-compost moisture and their relative quantities. Based on filter-cake moisture of 72%, recycled-compost moisture of 26%, and a recycle ratio of 1.5, an input moisture of 45% could be expected.

Moisture content of the recycled compost and filter cake was a major process limitation. Moisture content controlled particle size of the compost (Figure 4.2). This factor greatly affected process efficiency. Effective mixing within the composter was achieved only when the particle size of the compost was less than ⅛ inch. Mixing of recycled compost and filter cake was also affected in the same manner. Recycled compost of low moisture content (26%) and small particle size (⅛ inch) when combined with filter cake (72% moisture) produced a feed material easily mixed in the composter. If the recycled compost were insufficient and contained a high moisture content, the combination of recycled compost and filter cake would result in a sticky mass difficult to mix in the composter.

FIGURE 4.2: EFFECT OF MOISTURE ON COMPOST PARTICLE SIZE

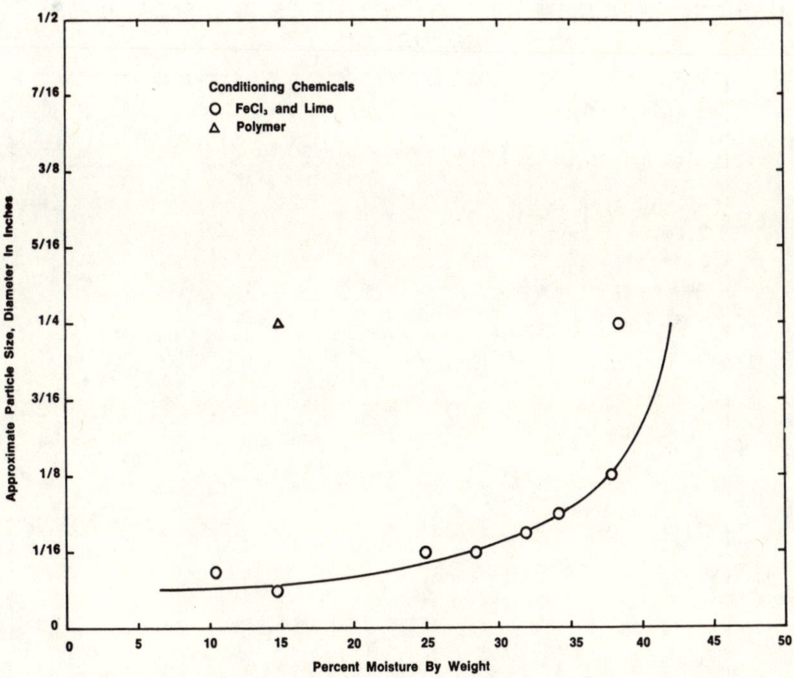

The biological action of the microorganisms present in the composter, as indicated by temperature, was affected by moisture. Activity was greatest at moisture contents of 20 to 30% and nearly nonexistent at moistures above 40% or below 15%. The extremes of moisture content, high and low, created physical problems related to the operation of the process. If the average moisture content of the compost mass exceeded 35%, the compost mass would begin to ball and clump together. This, of course, produced a mixing and aerating problem. Moisture content of the compost mass of 15% or less created a dusting problem because the forced aeration caused fine particles of compost to exhaust from the composter.

RECYCLING

Recycling of compost has several positive effects on this treatment process. The first, and probably most important, is the effect of the composter feed on the moisture content. If recycling were not used, the moisture content of the composter feed would be that of the sludge filter cake, and mixing problems (balling and packing) would develop. As was pointed out earlier, a feed moisture content of 45% or less can be obtained using a 1.5 recycle ratio when the filter cake has 72% moisture. When feed to the composter was consistently over 50% moisture, mixing problems soon developed.

FIGURE 4.3: EFFECT OF RETENTION TIME ON PERCENT SOLIDS REDUCTION

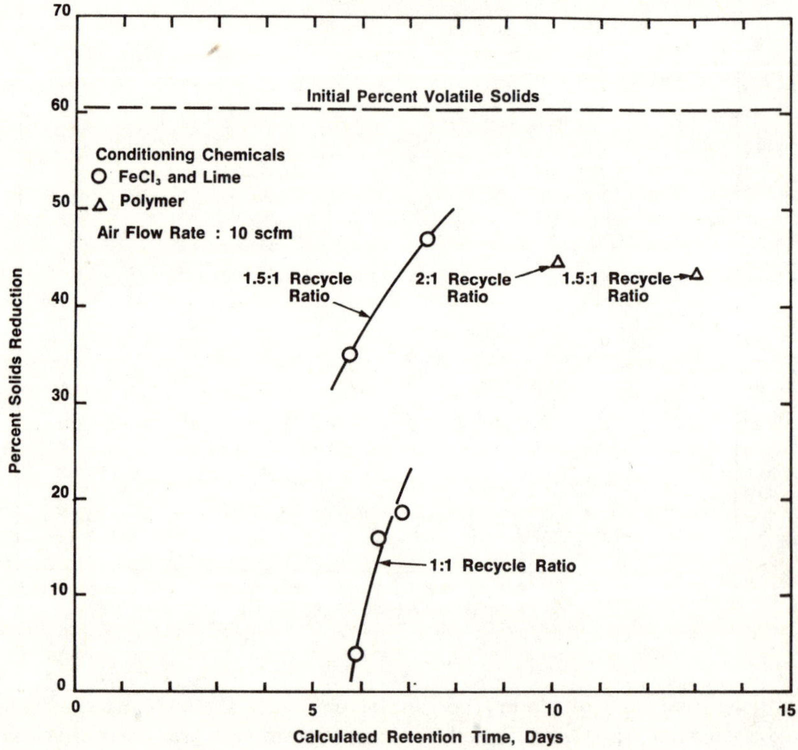

Another benefit of recycling is that the feed to the composter is thoroughly seeded with aerobic thermophilic microorganisms, and the biological activity is rapidly started, e.g., a 45% total solids reduction at 1.5 recycle ratio compared with an 18% total solids reduction at 1.0 recycle ratio, both at a 7-day retention time (based on the volume of dewatered sludge fed daily) (Figure 4.3). This indicates, then, that recycling compost, which seeds the filter cake and mixes the composter contents, produces beneficial effects.

The sludge filter cake conditioned with polymer indicated the same results. At a 2.0 recycle ratio, total solids were reduced 45%, with a 10-day retention time; at a 1.5 recycle ratio, they were reduced 43%, but with a 13-day retention time. The polymer-conditioned filter cake required more recycle than did the filter cake conditioned with lime and ferric chloride. This was because the polymer-conditioned filter cake contained 5 to 7% more moisture.

Recycling also increased the temperature of the feed from 10° to 20°F above that of the filter cake when a recycle ratio of 1.0 to 2.0 was used. This was believed to have a beneficial effect on the rapid response of the process.

When the sludge was conditioned with lime and ferric chloride, the filter cake pH averaged 11.0. The pH of the compost averaged 6.5, and therefore, it was probably beneficial to adjust the feed pH to be more compatible with the treatment pH by adding recycled compost to the feed.

PROCESS CAPACITY

The ability of composting to stabilize the dewatered sludge and reduce its volume and weight was of major interest in the test program. The process consisted of evaporation (drying) and biological destruction of volatile solids by thermophilic microorganisms. Drying is required not only for effective mixing, but because biological treatment is very slow to nonexistent at moisture contents above about 40%. Drying occurs in this process at a very rapid rate because of the high temperatures (140°F), continuous mixing, and aeration.

The moisture content of the dewatered sludge, which averaged 72%, was reduced to an average of 26% in the final compost. This represents a total average weight reduction of 62% and a moisture reduction of 87% by weight by drying alone. The percent water reduction is limited to 98% (15% final moisture content) by the dusting problem stated earlier. The volume reduction from the drying process because of shrinkage would probably account for more than 75% of the total volume reduction.

Destruction of volatile solids by biological action of thermophilic microorganisms caused a reduction of both weight and volume, but more important, stabilized the dewatered sludge and made it hygienic and free of viable plant seeds. This part of the treatment process is most important from the standpoint of final disposal. The final compost is relatively stable organically, is hygienic, and can be used as a soil conditioner or as land fill.

FIGURE 4.4: EFFECT OF RETENTION TIME ON PERCENT WATER REDUCTION

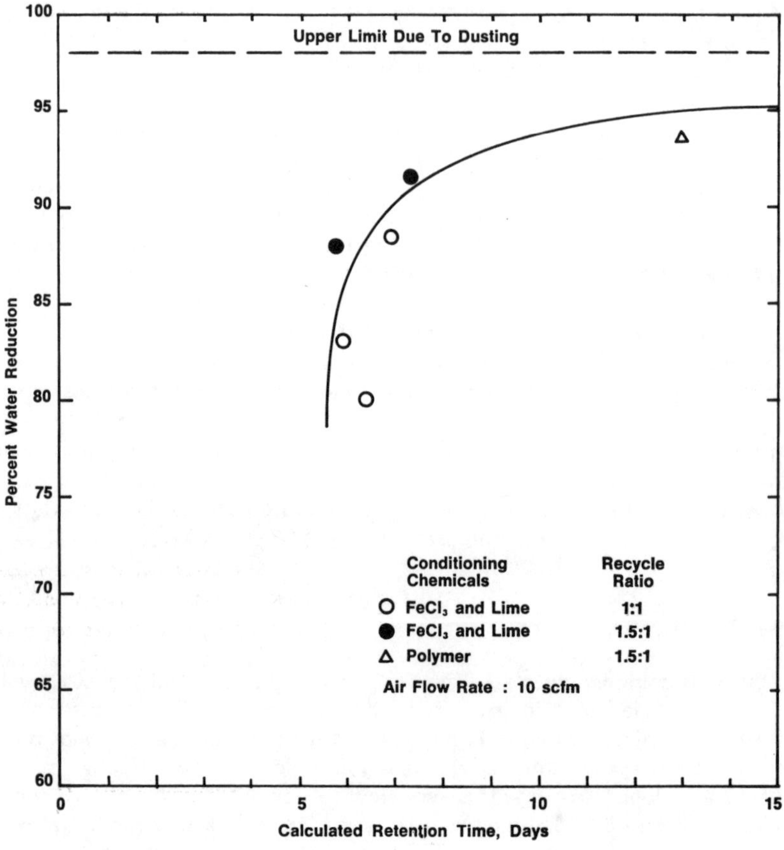

The intensity of the biological process was indicated by the high compost-bed temperature maintained with continuous mixing and aeration, the change in pH from 11.0 to 6.5, and the reduction of solids. An average total-solids (inerts included) reduction of 30% and an average volatile-solids reduction of nearly 50% occurred in the process. Solids reduction is limited to 61%, which is the volatile-solids content of the feed.

Again, the amount of volume reduction that results from biological action is not known. The average total volume reduction was 73%. All of the above average data were based on a retention time of 7 days. To obtain the calculated retention time, divide the total active volume of the composter by the volume of unpacked dewatered sewage sludge fed daily. The actual retention time in the composter is much greater than the time indicated in the above calculations because of the gross reduction of volume occurring during the process. Whereas the calculated retention time for the above averages was 7 days, the actual average retention time was closer to 26 days. The relation of calculated retention time to the moisture, weight, and volume reduction is based on a comparison of filter cake with compost (Figures 4.4, 4.5 and 4.6).

FIGURE 4.5: EFFECT OF RETENTION TIME ON PERCENT WEIGHT REDUCTION

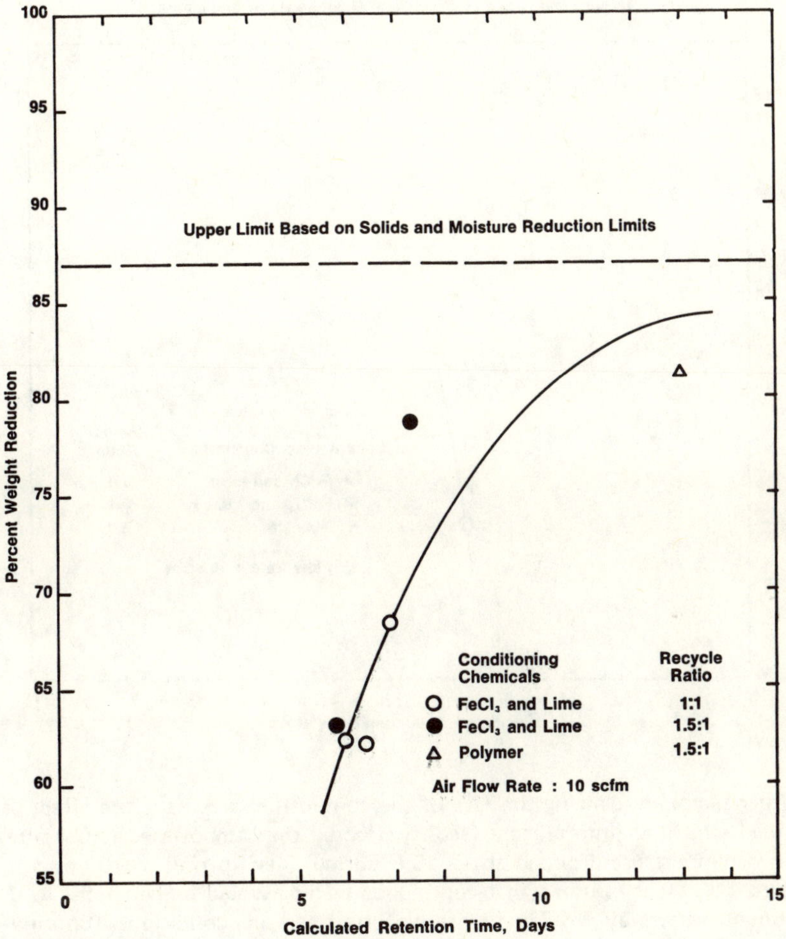

A flow diagram was made showing the fate of 100 lb of dewatered sewage sludge fed to the composter, with the amount of feed and products and their respective compositions (Figure 4.7).

FIGURE 4.6: EFFECT OF RETENTION TIME ON PERCENT VOLUME REDUCTION

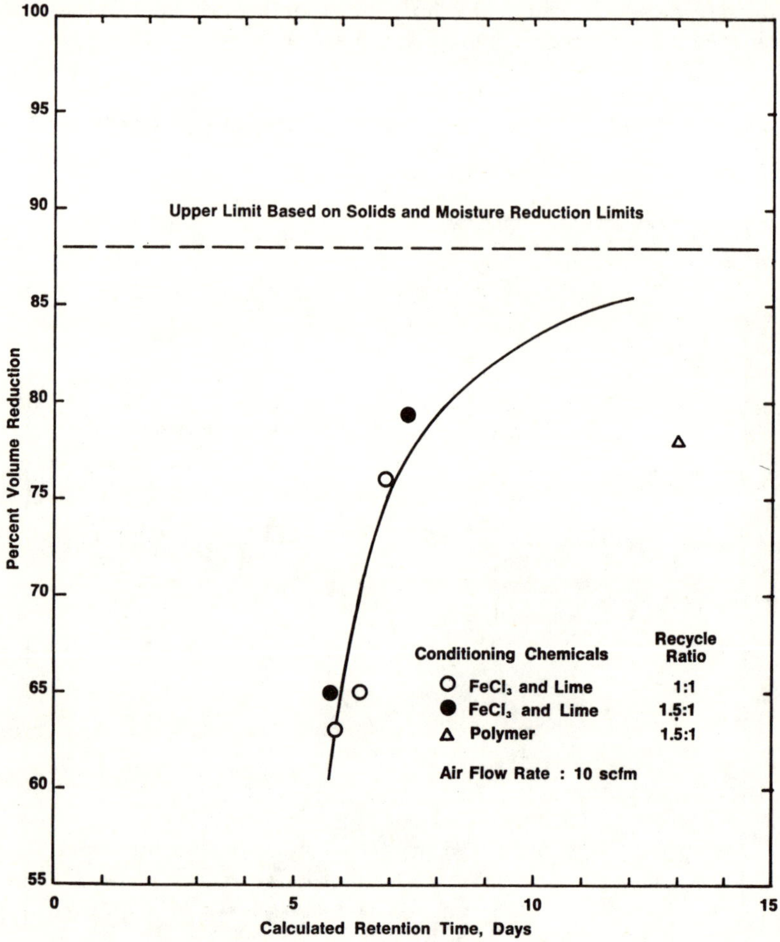

Another factor influencing the rate of the composting process is the effect of the feed schedule. Intermittent feeding affected the compost-bed temperature; the bed temperature dropped about 15°F during a feed period of 10 hours (Figure 4.8). It is assumed that continuous feeding would be beneficial to the treatment process by providing more uniform operating conditions. Continuous feed may significantly reduce the calculated retention time by allowing an increase in the amount of feed while still producing the results observed during intermittent feeding used during this study.

FIGURE 4.7: DEWATERED SEWAGE SLUDGE COMPOSTING – MATERIAL BALANCE

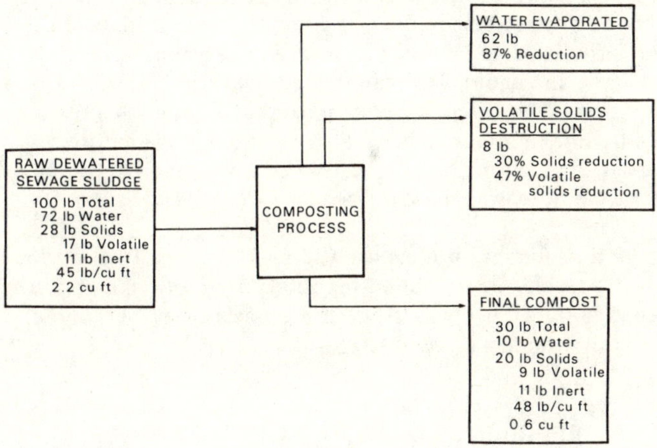

FIGURE 4.8: EFFECT OF INTERMITTENT FEEDING

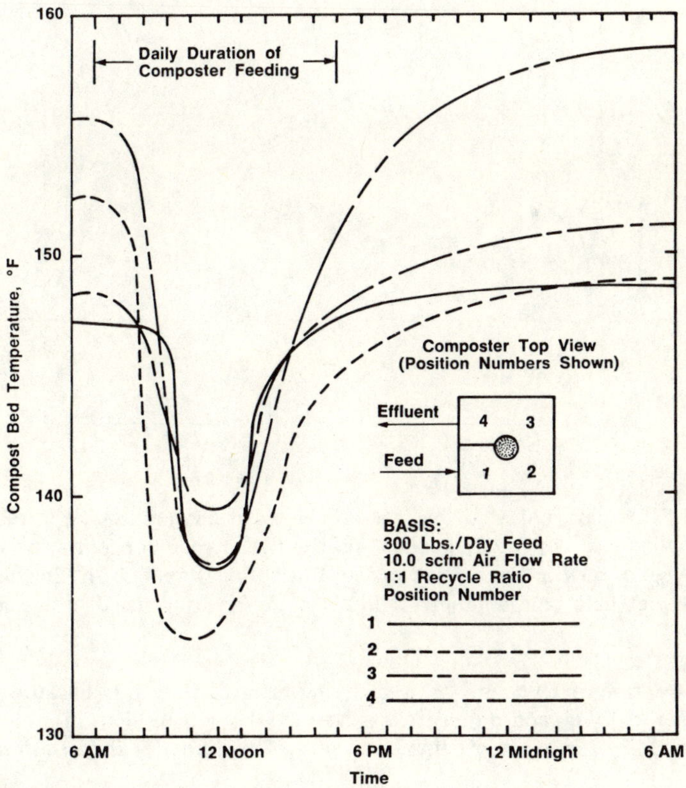

REQUIRED AIR

Forced aeration was applied to the compost bed and flow rates were measured. Aeration of the compost bed served two main functions. The first function was to provide a means for rapid evaporation of excess moisture fed to the composter. The large volume of air required for drying was shown to be well in excess of the air required to maintain aerobic conditions, the second function of aeration. This point was demonstrated (Figure 4.4); the treatment process required a minimum 4-day calculated retention time before solids destruction occurred. If feed to the composter had been predried, probably less retention time would have been required.

From the large reduction in moisture (80 to 90% by weight) during the process and the retention time required for solids destruction, composting of dewatered sewage sludge is obviously first a process of evaporation and second, of biological destruction of volatile solids.

FIGURE 4.9: EFFICIENCY OF OXYGEN UPTAKE RATE

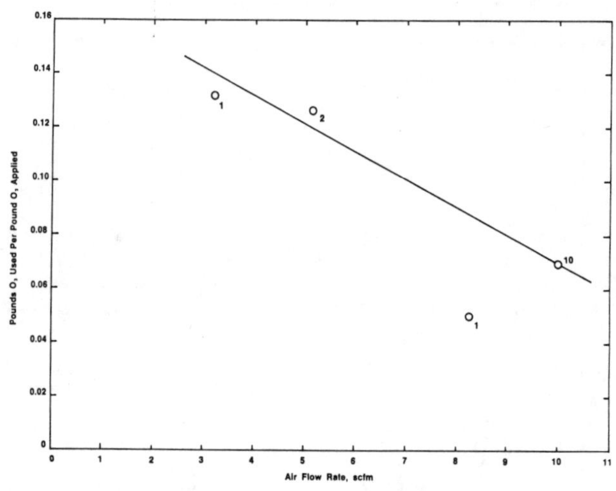

A straight line plot indicated that the efficiency of oxygen uptake in relation to air applied was not rate dependent on the the oxygen concentration present in the range of air-flow rates used during this study (Figure 4.9). In this figure numbers with data points indicate number of observations used to average each point.

An air-flow rate of 10 scfm was used during most of the study because it was sufficient for drying and did not excessively cool the bed. This rate represents a flow of 0.25 scfm per cu ft (based on 40-cu-ft volume) or 0.75 scfm per sq ft (based on 13.3-sq-ft area) to the compost bed. Forced aeration had a cooling effect on the compost bed (Figure 4.10). The temperature of the compost bed was recorded at four positions and three levels. (The position numbers are shown in a top view of the composter, and the levels noted indicate the distance from the bottom of the compost bed. A reading at the 6-inch level would be 6 inches upward from the bottom.) The three individual plots show that air rates up to 10.0 scfm did not lower the compost-bed temperature significantly. An average temperature decrease of 5°F was noted for air-flow rates between 5.3 and 10.0 scfm, and an additional 5°F decrease was observed for air-flow rates between 10.0 and 11.4 scfm.

FIGURE 4.10: EFFECT OF AIR FLOW RATE ON COMPOST BED TEMPERATURE

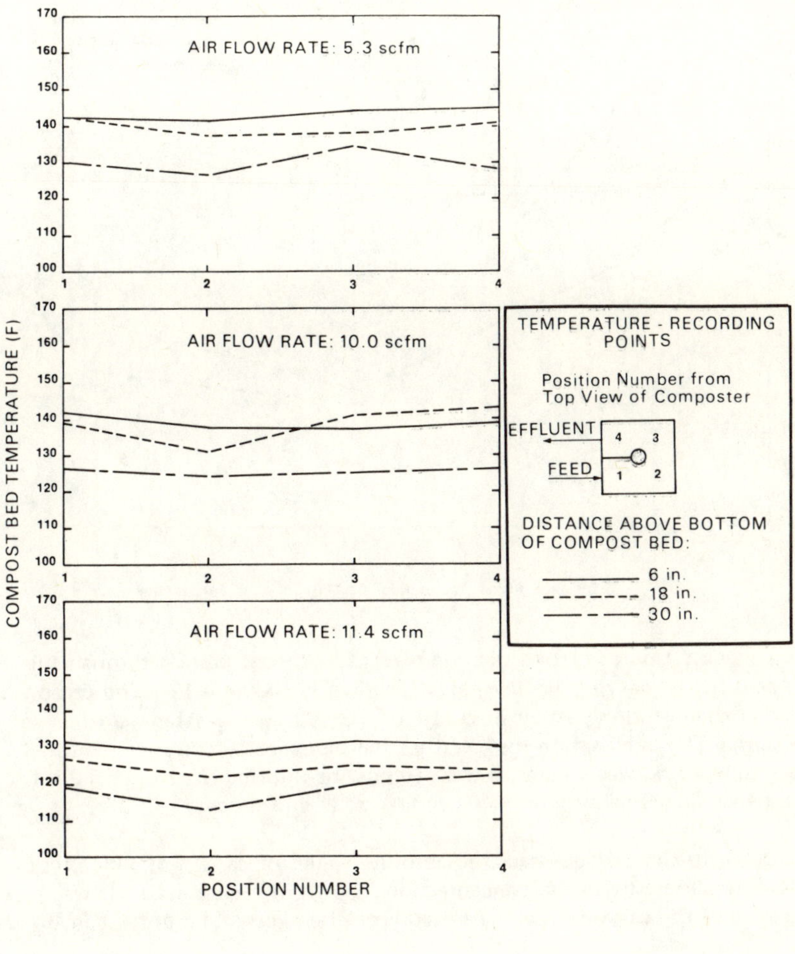

The concentrations of oxygen and carbon dioxide in volume percent as a function of compost-bed depth are illustrated (Figures 4.11 and 4.12, respectively). These samples were taken at the same sampling points as the temperature data discussed above. Even at low air-flow rates of 3 and 5 scfm, more than 18% oxygen by volume was present at all times (air usually contains 21% oxygen by volume). This again points to the fact that at the air-flow rates employed in this study, the oxygen concentration present was more than adequate at all times to maintain aerobic conditions. Also, as one would expect, the concentration of oxygen present in the air flowing through the bed was greatest at the highest air rate. The total depth of the bed was 36 inches.

FIGURE 4.11: EFFECT OF AIR FLOW RATE ON OXYGEN CONCENTRATION

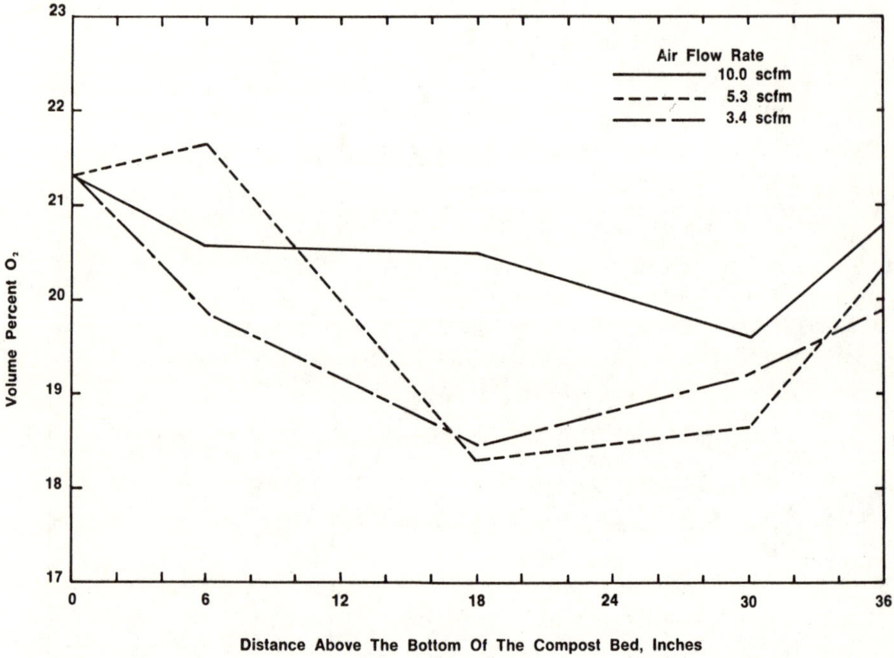

The concentration of carbon dioxide present in the air passing through the compost bed at varying air-flow rates is shown in Figure 4.12. The carbon dioxide concentration at an air-flow rate of 5.3 scfm was greater than that at 3.4 scfm. The explanation may be that the biological activity, as indicated by bed temperature, was greater in the compost bed during the run at 5.3 scfm and, hence, presumably more carbon dioxide was produced.

According to the oxygen used, the absolute values of carbon dioxide produced should be almost twice the concentrations shown in Figure 4.13. It was assumed that part of the carbon dioxide reacted with the excess lime present in the dewatered sewage sludge. Therefore, although the values recorded are not absolute, they do give an indication of process activity.

FIGURE 4.12: EFFECT OF AIR FLOW RATE ON CARBON DIOXIDE CONCENTRATION

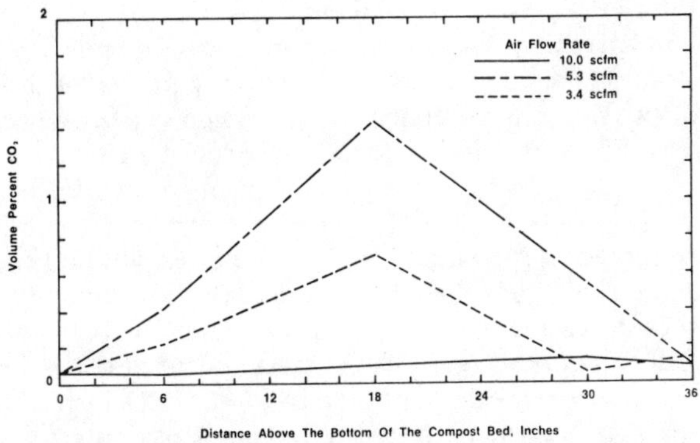

The lowest oxygen and highest carbon dioxide concentrations existed in the center of the compost bed. As fresh air fed to the bottom of the compost bed moves upward, oxygen is depleted and carbon dioxide is added. At a point somewhere between 18 and 30 inches (36-inch total bed depth) above the bottom of the compost bed, the effect of air diffusing downward from mechanical mixing occurred. From this point upward, the concentration of oxygen increased whereas the concentration of carbon dioxide decreased.

INFLUENCE OF CONDITIONING CHEMICALS

Conditioning chemicals used before vacuum filtration or any liquid-solids separation process may be classified into two main categories: inorganic chemicals, such as ferric chloride and lime, and polymeric organic chemicals. This variable in the composting of dewatered sewage sludge was studied to observe any differences in the process attributable to the conditioning chemicals. There has been a trend toward the use of polymers as sludge conditioners because they are easy to handle.

Different sludge conditioning agents produce filter cakes with differing physical properties. Sewage sludge conditioned with ferric chloride and lime produced a very firm filter cake that discharged from the filter medium easily. The polymers, however, produced a filter cake that was sticky and difficult to discharge from the filter medium. The filter cake conditioned with polymer also contained about 5% more moisture. Although an increase in moisture may account for some difficulty in discharge of the filter cake from the filter medium, it would not account for it becoming sticky. The above change in physical characteristics caused by the use of polymers resulted in many handling and mixing problems. Although the firm filter cake produced by ferric chloride and lime conditioning was easily conveyed by an inclined screw conveyor, polymer-conditioned filter cake produced a sticky, clumpy mass that was both difficult to convey and handle. The slight increase in moisture in the polymer filter cake may account for a portion of this problem, but does not totally explain it.

A similar problem was also encountered in mixing the recycled compost material with the filter cake. The filter cake produced with ferric chloride and lime, with its firm consistency, mixed easily with the recycled compost and produced a satisfactory feed to the composter unit. This easy mixing facilitated a well-seeded feed to the composter. Mixing the sticky, polymer-conditioned filter cake and recycled compost was difficult and produced a clumpy mass to be fed into the composter. This was then also difficult to mix in the composter.

The sludge-conditioning agents used also changed other physical properties of the compost bed. The filter cake conditioned with ferric chloride and lime resulted in a compost bed resembling sawdust in consistency. It mixed easily and posed no handling problems. The compost bed of dewatered sewage sludge conditioned with polymers had a particle size of gravel (1 inch or less). One result of this increase in particle size when polymers were used was that the air-flow rate through the compost bed was reduced from 10 to 3 scfm.

When the same air-flow rate was used with polymer as was used with ferric chloride and lime conditioning, the compost-bed temperature decreased by an average of 10°F. The change in particle size of the compost bed can only be accounted for by the change in chemicals used for conditioning because the compost bed contained the same or less moisture with the polymer than with ferric chloride and lime. This problem may be related to the bonding characteristics of the polymer as opposed to the powdery consistency of the inorganic chemicals used.

The final product and its disposal are also influenced by the conditioning chemicals. As stated before, the compost resulting from ferric chloride and lime filter cake was similar to sawdust in consistency. This material could easily be conveyed and disposed of in a variety of ways. The gravel-size material resulting from composting filter cake conditioned with polymers may have to be ground before final use or disposal. Other than these physical characteristics, the other properties of the final product were the same.

OFF-GASES

Off-gas samples were taken directly above the surface of the compost bed and analyzed. A typical gas chromatograph plot illustrates that the gas sample was split into well defined peaks and was easily measured (Figure 4.13). The composition of gases passing through the compost bed and the off-gas concentrations of oxygen and carbon dioxide changed in volume percent (Figures 4.11

and 4.12). Oxygen concentration, even at low air-flow rates, did not drop below 20% at the compost-bed surface (Figure 4.11). Carbon dioxide concentration did not exceed 0.1% at the compost-bed surface (Figure 4.12). The values of carbon dioxide and oxygen concentration in the off-gas should not be of concern.

FIGURE 4.13: GAS CHROMATOGRAPH SEPARATION OF COMPOSTER OFF-GASES

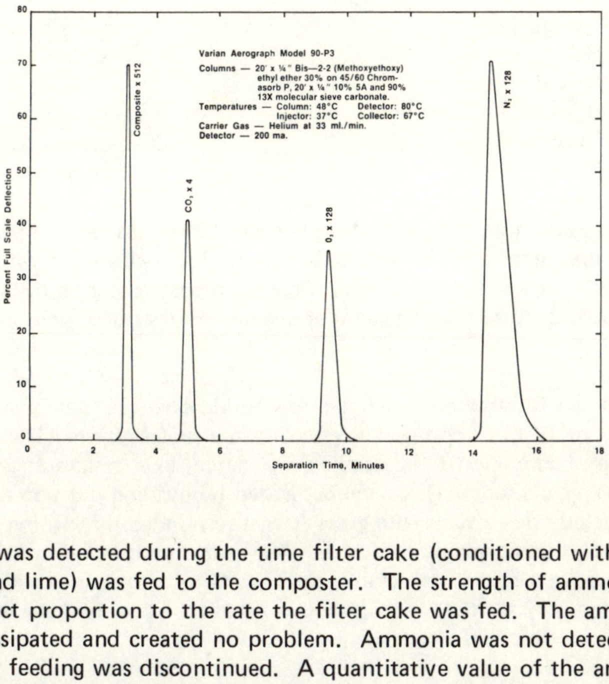

Ammonia was detected during the time filter cake (conditioned with ferric chloride and lime) was fed to the composter. The strength of ammonia odor was in direct proportion to the rate the filter cake was fed. The ammonia quickly dissipated and created no problem. Ammonia was not detectable 2 hours after feeding was discontinued. A quantitative value of the ammonia concentration was not measured with the gas chromatograph because the equipment would need to have been changed for ammonia analysis and other gas analyses could not then have been determined.

The applied air was believed to initially strip the ammonia (present because the pH was high when lime was used) from the compost. Although the gas chromatograph system was capable of detecting hydrogen sulfide, methane, and carbon monoxide, none of these gases was detected. These gases could have been present in trace amounts, but they were not detected by this analytical technique or by their odor. The odor of the combined off-gases resembled the odor of silage: a sweet, musty odor that might be encountered in a barn or a basement, not objectionable in any way.

CHEMICAL AND PHYSICAL PROPERTIES OF PRODUCT

The chemical and physical properties of the finished compost varied slightly with retention time. The following values (except for potassium and nitrate) are averages of 20 separate analyses:

Total Kjeldahl nitrogen (% N by weight)	2.21
Total nitrate (% NO_3^- by weight)	$<$0.01
Total phosphorus (% P_2O_5 by weight)	2.16
Total potassium (% K_2O by weight)	0.27
pH	6–6.5
Bulk density (lb/cu ft)	48.2
Moisture (% by weight)	20–35
Color	Dark brown
Odor	Musty

The fertilizer value (nitrogen and phosphorus) of the finished compost is low compared with commercial fertilizers, but is about equal that of cattle manure. More important is the ability of the finished compost to condition the soil to absorb and retain water. The general appearance is that of a rich humus such as peat moss.

The value of the finished compost as a soil conditioner was demonstrated by spreading 12 to 15 lb of compost per sq yd on a private lawn. The soil on which the finished compost was applied was sandy, low in humus, and low in natural fertilizer content. This compost added humus and did not cause any noticeable deleterious effects; the grass (lawn) responded by turning a rich, dark-green color and growing profusely. The added compost also aided soil aeration by preventing the soil from packing. All of the above characteristics have value and indicate that the finished compost used in this case was effective as a soil conditioner.

The moisture of the finished compost depended on retention time in the composter and reflected the degree of stabilization accomplished. The relation of compost temperature to number of curing days has been plotted for compost with moisture contents of 23 and 32% (Figure 4.14).

The temperature of the low-moisture (23%) finished compost decreased to ambient temperature almost immediately after being removed from the composter and placed in open piles. This indicates the completeness the overall process achieved while the compost was in the composter (25% solids reduction). After being removed from the composter, the high-moisture (32%) finished compost maintained its temperature for 5 to 6 days before decreasing to ambient temperature. This also reflected the degree of stabilization (4% solids reduction). These results indicate that the moisture of the finished product reflects the completeness of the overall treatment process and that both volatile-solids destruction and drying vary with retention time. A sample of the finished compost was cultivated for 1 month under controlled laboratory conditions to determine if viable plant seeds were present. The sample was watered daily with distilled water and left in a warm, lighted area. After 1 month, the sample was examined to determine if seeds, possibly present in the sewage sludge initially, were viable after composting. No seeds germinated; the compost retained water well and did not pack or lump. The odor was musty but not objectionable. This test, although conducted only once, added valuable information about the characteristics of the finished compost.

Particle size of the final product also varied with the retention time because size was determined to some extent by moisture content. Moisture contents of the finished compost (between 20 to 28%) resulted in a fine-textured material similar to sawdust with a bulk density of about 48 lb per cu ft.

FIGURE 4.14: TIME-TEMPERATURE RELATIONSHIP IN COMPOST CURING PROCESS

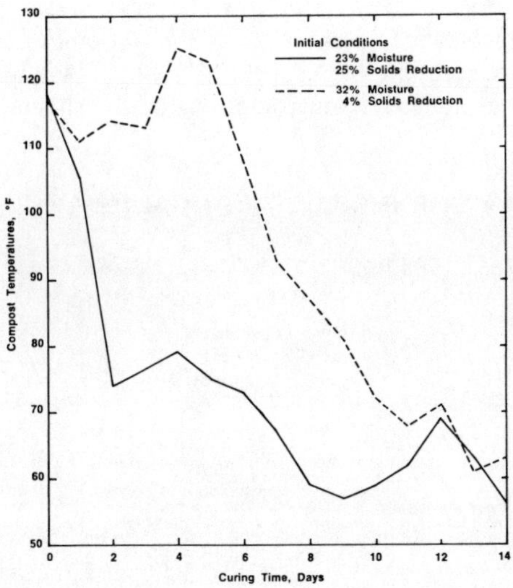

The pH level of the end product was consistently between 6.0 and 6.5. This level should pose no problem in handling, final use, or disposal. The concentration and type of organisms remaining in the finished compost would make it a hygienically safe soil conditioner. Pathogenic organisms initially present and indicator organisms added in the pathogenic study were destroyed in a maximum of 3 days. The excellent destruction of pathogens was probably due to the more or less uniformly high temperature maintained and the thorough mixing of the compost bed. The thermophilic organisms, none of which were pathogenic, should pose no problem at ambient temperatures.

In general, the physical and chemical properties of the finished compost would make it a safe soil conditioner with some fertilizer value.

SURVIVAL OF PATHOGENIC ORGANISMS

Because of the possible uses of the finished compost, a very important phase of this study was the survival of pathogenic organisms through the treatment process. Studies of the destruction of pathogens in composed garbage and a combination of sewage sludge and garbage indicate that destruction occurs primarily as a result of two actions: thermal kill by high temperature and the effect of antibiotic action. This study was made to determine the destruction of pathogenic type organisms associated with sewage and sewage sludges.

Basically, the organisms chosen were those that simulate actual conditions and that also cause infection when entering the body via the gastrointestinal route. Because Clostridium and *Staphylococcus aureus* do not meet these criteria, they were not used. The choice was also based on the anticipated problems of detection in the filter cake and final compost, their possible presence in sewage and sewage sludges, and the various types (bacteria, virus, etc.) that could be present. The four pathogenic-type organisms studies were *Salmonella newport*, a bacterium; *Candida albicans*, a fungus; *Ascaris lumbricoides*, a nematode (metazoan parasite ova); and poliovirus Type I, a virus. A flow diagram of the sequence of testing and method of microbial analyses is shown in Figure 4.15.

FIGURE 4.15: ASSAY FOR INDICATOR ORGANISMS

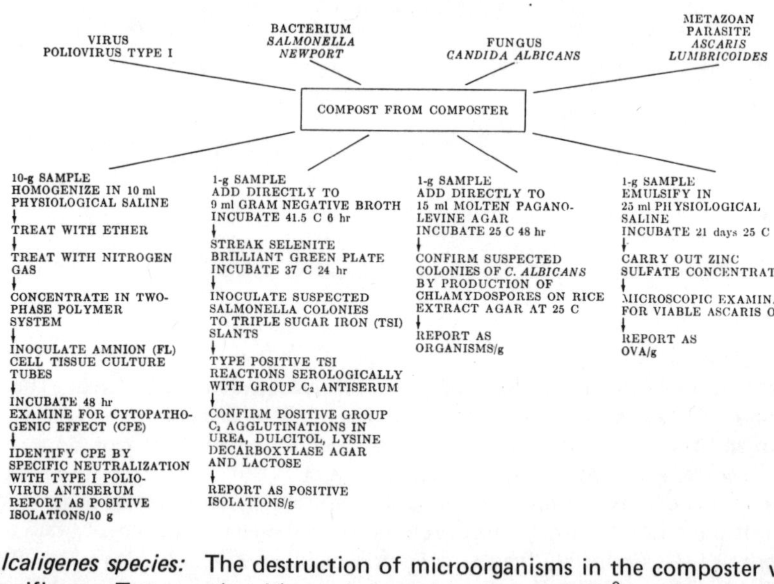

Alcaligenes species: The destruction of microorganisms in the composter was significant. Tests, made with nutrient agar incubated at 37°C, indicate that total organisms were reduced from 2.1 x 10^7 to 1.1 x 10^3 , or 99.99% kill. The count of gram negative enteric bacteria, employing MacConkey plates for detection, showed 5.6 x 10^5 organisms per gram in the filter cake and zero organisms in the final compost. The predominant organisms in both the filter cake and final compost were Alcaligenes species. These are nonpathogenic, alkali-producing bacteria associated with the breakdown of proteins in sewage. Interestingly, tests indicated a 50-fold increase in thermophilic organisms from 2 x 10^4 organisms per gram in the filter cake to 1.5 x 10^6 organisms per gram in the final compost.

Laboratory and insertion tests were made with the use of each of the microorganisms indicated in Figure 4.15. Aseptic sampling techniques were used throughout these studies of pathogens. The results for each are discussed separately.

Salmonella newport: These pathogenic bacteria are found in sewage. When laboratory tests were made to determine the thermal death point (TDP) (time is a constant and temperature is varied) and the thermal death time (TDT) (temperature is a constant and time is varied) of *S. newport*, the TDT was 30 min at 60°C and the TDP was 65°C in 30 min. These results indicated that *S. newport* is susceptible to kill at temperatures existing during composting treatment (60°C or 140°F). This organism was not isolated in either the filter cake or final compost before inoculation. Insertion studies involved 117 microbial analyses after the compost in the composter (40 ft^3 volume) was inoculated with 2.7 x 10^{12} organisms.

Samples of the compost were removed from the inlet and outlet portions of the composter at given time intervals. One hour after inoculation the tests indicated *S. newport* was present at the inlet zone and after 4 hr, at the outlet. No *S. newport* was found in the inlet or outlet zones 25 hr after inoculation. Sampling continued for 10 days with no further isolation of any organisms, not even coliform, on the selective medium or on selenite brilliant green agar.

Candida albicans: These pathogenic yeasts survived well at 37°C in both water and sewage. The selective medium used to isolate this organism was Pagano-Levin medium. Again, no organisms were isolated from the filter cake or final compost before inoculation. The TDP was established at 80°C for 30 min, and the TDT was 60 min at 70°C. This also showed susceptibility to kill at composting temperatures.

Insertion studies employing 2.0 x 10^9 organisms mixed into the composter feed were tried. One hour after the compost had been inoculated with the microorganism, analysis of samples taken at the composter inlet showed isolation of *C. albicans* too numerous to count. At least one organism per gram was isolated after 28 hr at the outlet from the composter. No further isolation occurred during the next 3 days of sampling involving 30 additional analyses.

Ascaris lumbricoides: These pathogenic nematodes (small worms) are found in the intestines of hogs. The ova used during this study were taken from hog intestines obtained from a local abattoir. After the whole worm was removed from the hog intestines, their uteri were removed, homogenized, placed in a saline solution, and refrigerated for future use. Laboratory tests indicated a TDP of 60°C for 60 min. No tests were made on the TDT.

Insertion studies involving the inoculation of 1.4×10^8 organisms into the composter were performed. The results of these tests showed that viable ova were not present 4 hr after inoculation. Ova were found 1 hr after inoculation at the inlet and up to 76 hr at the outlet of the composter; after 76 hr, ova were not detected. The ova began to disintegrate 24 hr after inoculation; this accounts for their total absence after 76 hr. A total of 110 microbial analyses were made.

Poliovirus Type I: This virus was the most susceptible to high-heat kill; its TDP was 50°C for 30 min and its TDT was 5 min at 60°C. The separation technique used to isolate this virus involved a two-phase polymer system to concentrate the samples, followed by inoculation into cultures of FL cells and examination for cytopathogenic effects. The final test consisted of treating suspected samples with antipoliovirus Type I antiserum and observing the results.

Initial analysis of the sewage sludge and compost revealed no virus of this strain was present. In the insertion studies, 2×10^7 organisms were mixed into the composter feed. All samples taken at the inlet and at the outlet of the composter were negative. This is reasonable since the TDT of the virus is only 5 minutes at 60°C.

DESIGN CRITERIA

Application of the composting process requires that certain basic criteria be known. In general, the results of this study established these basic design parameters, and these parameters structured the design of a compost process for a primary and secondary sewage treatment plant with a sewage flow of 1 mgd. When the sources and quantities of the sludges to be treated were listed, it was noted that the secondary portion of treatment adds about 30% more sludge to be processed (Table 4.1).

Although the test work involved only the combination of primary and secondary sludges, there is no reason to believe that the process will not work just as well on primary sludge alone. Based on the results obtained from the operation of the pilot plant, a calculated retention time of 7 days (actual retention time about 26 days), an air-flow rate of 0.25 scfm per cu ft and a mixing horsepower of about 0.05 hp per cu ft (based on paddle mixing, which is not necessarily recommended) is suggested. The size of the equipment required to process the volume of sludges for the 1-mgd plant varies according to the treatment phase (Table 4.2).

In these design criteria, the compost depth is set at 3 ft. Although this parameter was not studied, it was believed that lesser depths would begin to cause increased power for mixing. Mixing appeared to be a significant parameter in the process efficiency.

TABLE 4.1: SOURCES AND QUANTITIES OF SOLIDS TO BE TREATED

Item	Primary treatment	Primary and secondary treatment
Raw waste:		
SS* (mg/liter)	250	250
BOD_5† (mg/liter)	200	200
Primary treatment:		
65 percent SS removal (mg/liter)	160	160
40 percent BOD_5 removal (mg/liter)	80	80
Secondary treatment:		
85 percent removal of BOD_5 from primary effluent (mg/liter)		100
BOD_5 removed, converted to solids (~50 percent) (mg/liter)		50
Total dry solids removed (mg/liter)	160	210
Total dry solids in 1 mgd (lb)	1, 340	1, 750
Total wet solids (~70 percent moisture) in 1 mgd (lb)‡	5, 100	6, 700
Total wet solids (45 lb/cu ft) in 1 mgd (cu ft)	114	149

*Suspended solids.
†Biochemical oxygen demand, 5 day.
‡Figures adjusted for the addition of 15 percent conditioning chemicals on a dry solids basis.

TABLE 4.2: DESIGN CRITERIA

Item	Primary treatment	Primary and secondary treatment
Composter:		
Volume (cu ft)	800	1, 000
Depth (ft)	3	3
Mixing, using paddles, (hp)	~40	~50
Air required:		
Volume (scfm)	200	250
Pressure (psig)	1	1
Horsepower	~2	~2

The above design does not include the appurtenant facilities required to feed, recycle, mix, and discharge the final product. The operation of the pilot plant indicated that the entire process could be nearly automated, with a minimum of manual operation required. Temperature and dissolved oxygen probes could be used to control the rate of mixing and air-flow rate. Moisture sensors in the compost bed and feed could be used to control recycle and feed rate, and a bed level control could be used to control the discharge of final product.

The types of equipment for materials handling would necessarily be designed on the basis of volume, consistency, and location requirements of the composter layout.

CONCLUSIONS

The process was capable of reducing moisture, volume, weight, and solids by an average of 87, 73, 73, and 30%, respectively. Continuous mixing and aeration of the compost bed was required to ensure good odor-free treatment. The air requirement for drying was critical and well in excess of the air required to maintain aerobic conditions.

Recycling was not only beneficial, but necessary; it facilitated moisture adjustment of the feed and microbial (aerobic thermophilic) seeding. The final compost had a nitrogen and phosphorus content of a little over 2% each, expressed as N and P_2O_5; this compost worked well as a soil conditioner and was found to be free of indicator pathogens and viable plant seeds.

The study of pathogens indicated that all of the test organisms (*Salmonella newport, Candida albicans, Ascaris ova,* and poliovirus Type I) inserted in the compost were killed in less than 3 days' retention in the composter.

The chemicals used had a dramatic effect on the process. Sludge conditioned with lime and ferric chloride produced compost that mixed and handled better than sludge conditioned with polymer. Moisture content of the feed and the compost bed were critical for proper treatment. The optimum moisture content in the compost bed was 25 to 30% by weight.

Cellulose Degradation in Composting

Recently there has been an increasing interest in composting. However, over the past 30 years the character of household refuse has changed. The quantity of paper and paper products discarded has increased drastically thus making solid waste more resistant to biological treatment. Presently municipal refuse typically consists of from 50 to 80% paper.

During the first 2½ years, the Manhattan College solid waste project concerned the composting of a synthetic waste composed of 100% paper, a cellulolytic material. It was the degradation of cellulose, the major component of paper, which was considered to be the rate limiting step in composting. This earlier work is discussed in the light of more recent results.

It was found during the earlier work that cellulose degrades slowly and it was concluded that the significant improvements required which would result in an economical composting process would have to be of such a magnitude that it was beyond optimization of cellulose alone. For this reason, in January 1969, the direction of the research was altered to determining the environmental conditions necessary for optimum degradation of typical municipal refuse. This research is described in PB 215,722 (1973) by R. Regan, J.S. Jeris, R. Gasser, K. McCann and J. Hudek from which the material in this chapter is derived.

LABORATORY PROCEDURE AND EQUIPMENT

Composting investigations were performed using three major experimental systems: (a) bench scale composters, (b) shake flask apparatus, and (c) Warburg respirometer. The continuously fed experiments were made using three bench scale composters. Batch fed studies were carried out using the shake flask apparatus, and the Warburg respirometer. Each of these apparatuses allowed for close control of the desired environmental parameters and was sufficiently flexible to operate over wide ranges of conditions. The general procedure used for obtaining results during the research is outlined in Figure 5.1. The research was carried on in three major steps.

Step 1 – Samples of waste material were placed into the battery of bench scale composters. The composters were operated until the microorganisms present in the waste were assumed to have become acclimated at certain desired conditions. Significant quantities of composted material to be used as seed were produced in this manner.

Step 2 – Wide ranges of chemical and physical parameters were investigated in batch fed experiments using the shake flask and Warburg apparatus. Samples of waste material were mixed with significant amounts of seed as prepared in Step 1 to minimize lag phases in obtaining preliminary experimental results.

Step 3 – Using the most promising results of Step 2, narrower ranges of the various study parameters were extensively examined in the bench scale composters on a continuous feed basis. In this manner, the optimum ranges were defined for various environmental factors in composting typical mixed refuse.

FIGURE 5.1: OUTLINE OF PROCEDURE FOR OBTAINING EXPERIMENTAL RESULTS

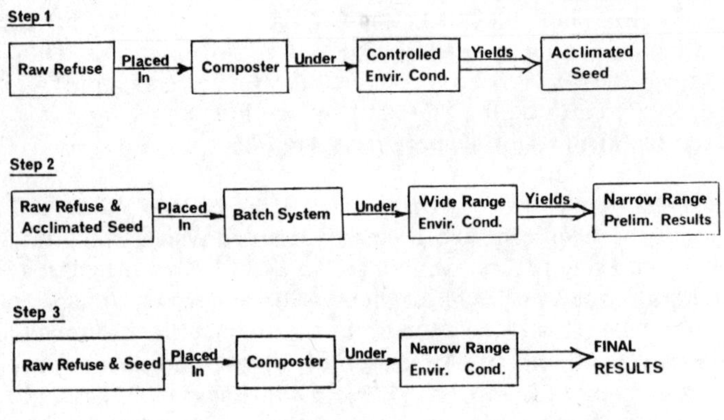

Source: PB 215,722, 1973

Bench Scale Composters

A schematic drawing of the experimental bench scale composting system is shown in Figure 5.2. Each of the three composters used was a horizontal stainless steel cylinder, 12 inches in diameter, 2 ft long with a total volume of 1.5 ft^3.

FIGURE 5.2: LABORATORY COMPOSTER SYSTEM

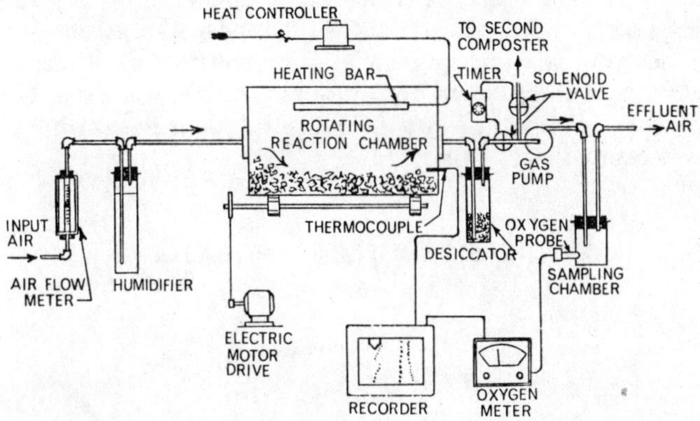

Source: PB 215,722; 1973

Under operating conditions, an 8 inch port on the wall of each cylinder made access convenient for loading, sampling and cleaning. Waste material tended to adhere to the walls during operation. The walls were scraped daily by hand and the material manually mixed with the remaining solids. Each composter contained a total weight of 1,000 to 3,000 g and filled about one-half the cylinder volume.

Continuous heat was supplied by an electric heating rod suspended in the air space directly over the composting mass. The temperature was maintained at a constant level by controlling the power to the heating rod with a rheostat. Excess heat losses were reduced by placing the battery of composters in a walk-in type incubator maintained at 37°C and by using insulation around each cylinder. In this manner, the control of the temperature was not dependent on microbial activity. The temperature could be adjusted to any desired level and used as a controlled variable for studying its effect on composting efficiency.

Mixing and aeration of the material was provided by intermittent rotation of the cylinder. During these studies the drive motor was activated to allow one rotation of each cylinder for time cycles of 30 minutes to 3 hours.

Air was supplied to each unit through a rotometer at a rate sufficient to maintain an oxygen residual of at least 10% in the effluent gas. Continuous analysis of the residual oxygen of each unit was measured automatically every 4 hours by an oxygen probe placed in the sampling chamber and monitored by a multipoint strip chart recorder. Daily samples of gas were also taken and analyzed by gas chromatography to determine the carbon dioxide content of the effluent.

Shake Flask Apparatus

The shake flask apparatus was composed of four incubators each containing eight shake flasks. The flasks were mounted on the arms of two Burrell wrist-action shakers which were located between the incubators as shown in Figure 5.3. Each incubator was 25" x 20" x 16" and constructed of masonite with 2 inches of rock wool insulation on five sides. The sixth side was an observation door. A heating rod mounted in the incubator provided heat which was regulated by a rheostat.

FIGURE 5.3: SHAKE FLASK COMPOST APPARATUS

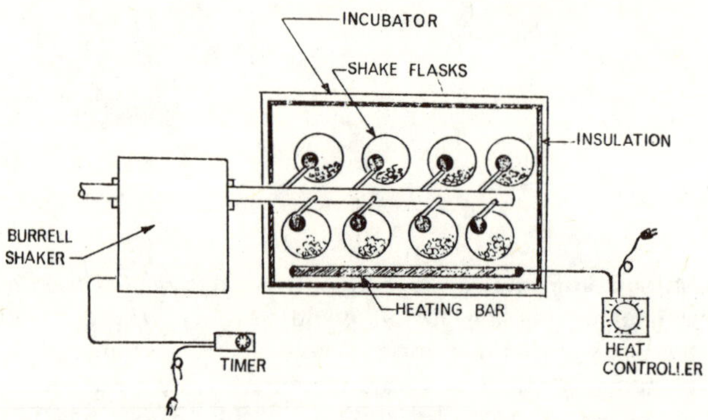

Source: PB 215,722; 1973

The flasks used were 500 ml Erlenmeyers and sealed with a rubber stopper containing a glass tube which served as a gas sampling port. Approximately every 3 to 4 hours during a run, a gas sample was taken from each flask, for analysis and the flask was completely aerated. The Burrell shaker provided a tumbling action for a 2 minute period every 2 hours. A maximum of 32 flasks, at four different temperatures, could be set up at one time ensuring coverage of a wide range of conditions.

To initiate a shake flask study, acclimated seed was obtained from the bench scale composters and placed in each flask together with refuse, nutrients and additives to produce the desired conditions. Duplicate flasks were normally set up for each study condition. Controls were used to compare the activity of the seed material with those containing refuse and other test additives. As a run progressed, nutrients and additional substrate were added as necessary. Generally, a run was concluded after one week of data collection.

Warburg Respirometer

The Warburg Respirometer was used to determine the total oxygen consumption and oxygen consumption rates of the various solid wastes used in the investigations. The Warburg apparatus shown in Figure 5.4, measured manometrically changes in the gaseous content within the closed system, containing the biologically active mass. As oxygen was consumed by the aerobic microorganisms, the carbon dioxide produced was absorbed by potassium hydroxide within the center well of the flask.

The resulting decrease in pressure within the closed system is related to the oxygen consumed by the microorganisms present. This type of apparatus is ideally suited for composting studies, in which a high degree of environmental control and a wide range of conditions require study. In the apparatus, 18 flasks, or separate study systems could be set up at one time. A sample of material was placed in a flask at a specified moisture and temperature level. Since each flask was opened only periodically for reaeration, evaporation losses from the sample were minimized for the duration of the test period.

FIGURE 5.4: WARBURG APPARATUS

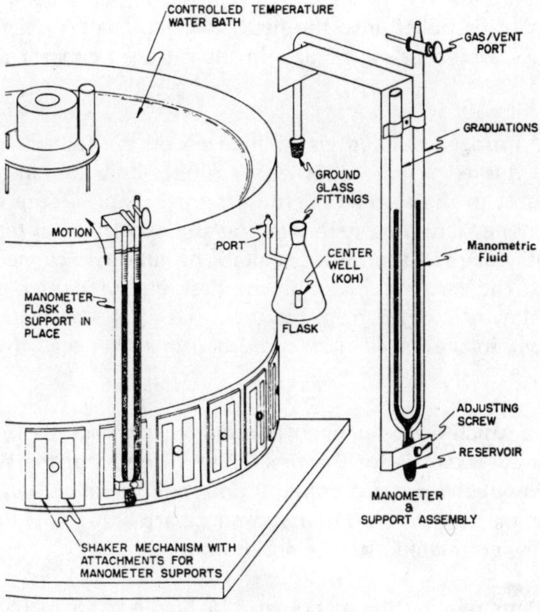

Source: PB 215,722; 1973

The basic operating procedure for application of the Warburg respirometer to the composting research was as follows. Approximately 100 g of material containing desired amounts of mixed refuse and acclimated seed (both on a dry basis), from a continuously fed laboratory composter operating near the appropriate condition, were blended briefly at a slow speed to insure a homogeneous mixture. Sufficient ammonium chloride and potassium phosphate were added prior to blending to provide the required nutrients.

The initial pH and moisture level of the mass were determined. The sample was divided into smaller portions in which the equivalent dry weight used in each flask was 1.4 g. Sufficient water and chemicals were added to each portion to attain the desired experimental conditions.

Duplicate samples, at a given set of conditions, were placed in 125 ml Warburg flasks. In the center well of each flask 1 ml of 20% potassium hydroxide was placed for carbon dioxide absorption. The flasks were attached to mercury filled manometers and placed in the constant temperature bath at 50.8° or 59.6°C. The Warburg apparatus was placed in motion and the flasks were allowed to reach a constant temperature with the system open to the atmosphere for at least 10 minutes. The flasks were sealed. Manometer readings were made.

Because of the great oxygen demand caused by the large amount of organic matter present in the sample, special reaeration procedures were required during the run. If the vacuum was released in the flask by simply opening the vent and allowing air to be pulled into the flask, the result, after relieving the pressure several times, would be an increase in the nitrogen concentration in the system.

The problem of nitrogen buildup was minimized by the following procedure. The flasks were flushed several times with a rubber bulb so that more oxygen was made available to the aerobic microorganisms. A procedure was developed in which oxygen was saturated with water at the experimental temperature and placed in a tube. As reaeration was required the tube was connected to the Warburg flasks. The vacuum created in the flask due to aerobic activity would be relieved by flow of oxygen from the tube. Using this system, the oxygen concentration within the sealed flask could be no greater than that in atmospheric air.

The manometer, which gives the change in pressure in each flask, was measured at 15 to 30 minute intervals for the first 12 to 14 hours of the Warburg run. Typically the cumulative oxygen consumption data for these systems followed a linear rate for the first 8 to 10 hours; the rate after this time decreased. The Warburg run was terminated after 24 to 28 hours.

The flask constants relating the manometric deflections for mercury to the oxygen consumption were calculated at 50.8° and 59.6°C to be approximately 0.15 to 0.16 mg O_2/g dry mass when using samples containing a total of 1.4 g dry mass. The oxygen uptake rate was determined by measuring the rate of change of the total oxygen consumption for the first eight to ten hours of the run. The results were correlated to the various environmental factors imposed on each sample.

In some of the Warburg studies, samples were removed during the first 12 to 14 hours of the run to determine the carbon dioxide production rate. The carbon dioxide concentration of the potassium hydroxide in the center well of the flask was measured using a Beckman carbon analyzer, Model No. 915. The carbon dioxide measured was assumed to be linearly produced during the period of linear oxygen production.

PREPARATION AND ANALYSIS OF SOLID WASTE AND SEED

Solid Waste Substrate

The solid waste used as feed for the research included newsprint, typical mixed refuse containing 60 to 70% paper and composted mixed refuse from Altoona, Pennsylvania. The cellulose content was significant in each of these substrates.

Two methods were employed for obtaining mixed refuse. Initially, batches of urban refuse were collected biweekly and stored in metal containers until needed for experimental use. This method was inadequate because the chemical characteristics of the refuse varied from batch to batch. Also the refuse deteriorated during storage making comparison of the experimental results difficult. To improve the degree of control, a synthetic refuse, with characteristics similar to municipal solid waste was prepared in the laboratory. It consisted of wastes from the college cafeteria, supermarket wastes and newsprint. Typically meat and fish scraps, lettuce and other vegetable trimmings were used.

The solid waste feed was prepared by cutting and grinding the waste materials with a garden shredder. Materials were cut to a particle size no greater than one-quarter of an inch. Since the waste materials contained insufficient nutrients, nitrogen in the form of ammonium chloride was added. The wastes were also supplemented with phosphorus. The moisture content was adjusted to suitable levels for composting with tap water. It was assumed that the tap water supplied necessary trace elements.

Seed

Purified cultures growing on cellulose in liquid media were used as seed for the startup of some of the composting studies involving newsprint. Other experiments using newsprint were performed using a maximum of 10% composted mixed refuse as seed. Batch fed studies were made on composted refuse, as it was received. No additional seed material was used, since suitable microorganisms were assumed to have remained in the composted waste. The use of high recycle in the studies involving mixed refuse in batch and continuously fed composters made use of supplemental seed material unnecessary.

Routine Analyses and Measurements

In addition to the continuous record of residual oxygen concentrations of the effluent gas of each of the bench scale composters, other physical and chemical parameters were measured routinely. The measurements made during active composting included mass temperature, air supply rate, carbon dioxide concentration, pH, moisture content, bulkweight and gross weight.

Analyses were performed to determine the chemical composition of each waste material before and after composting. The number and types of chemical analyses made on each waste were varied due to the nature of the material. For example, a greater number of measurements were required to characterize mixed refuse than were needed for newsprint.

Overall, the chemical analysis used to typify each waste material included ash, volatile matter, crude fiber, lipids, bulkweight, ammonia and total nitrogen. Carbon dioxide concentration of the effluent gases were determined using an Aerograph gas chromatograph with a silica gel column.

Specific Analyses

Concentration of Leachate: The chemical oxygen demand (COD) of refuse which is related to the concentration of organic matter present has been suggested as a parameter for use in composting. In the present research, the COD of the leachate was used to indicate the degree of stabilization by composting.

Part of the organic matter present in mixed refuse is known to be or become soluble. It was believed that if reliable and consistent COD values could be obtained for the soluble matter leached from refuse and composting mass, it would be possible to measure the effectiveness of the stabilization by comparing the initial and final COD of the soluble fractions.

The following procedure was used for determining the leachate COD of mixed refuse. Samples of the mixed refuse and product from the bench scale composters were taken for analysis. Four grams of moist mass (approximately 65% moisture) were weighed using a triple beam balance accurate to ± 0.1 gram. These samples were then diluted with 500 ml distilled water and mixed by means of a magnetic stirrer for one hour. After this time, the samples were filtered through a fine mesh stainless steel screen held on a California Buchner funnel. The supernatant was centrifuged and the soluble fraction analyzed for COD by a rapid procedure.

A comparison of the COD concentration of this soluble matter before and after composting gave a measure of composting efficiency and the degree of refuse stabilization. Also, for experimental control it was possible to adjust the mass turnover time in the bench scale composter to study the rate of removal of the soluble fraction from the refuse.

Organic Carbon: Organic carbon of the mixed refuse and composted matter was determined using a Beckman Model 915 Total Organic Carbon analyzer. Measurements of total carbon and inorganic carbon were made and their difference gave the organic carbon content. The analyses were based on conversion of sample carbon into carbon dioxide for measurement by the Beckman nondispersive infrared analyzer, Model 215A.

The samples required for the analysis were prepared using the crude fiber test which includes hydrolysis with acid and base. The analyses were performed on the residue remaining on the stainless steel screen after washing as described previously in Concentration of Leachate. Hydrolysis consisted of two 20 min mixing periods; one in 100 ml of 0.1 N sodium hydroxide, followed by one in 100 ml of 0.1 N sulfuric acid. The acidic and basic filtrate was neutralized, diluted with distilled water to a volume of 500 ml prior to injection into the carbon analyzer. The sum of the quantities of organic carbon measured were correlated to the weight loss of samples.

Volatile Acids: The volatile acid concentration of mixed refuse and composted mass was determined by the column partition method using silicic acid, as outlined in *Standard Methods.*

Solid Waste Characteristics

Table 5.1 summarizes typical values obtained from the routine analyses performed on solid waste materials used as feed during the duration of the project.

TABLE 5.1: SOLID WASTE CHARACTERISTICS

	Newsprint	Altoona Compost	Municipal Refuse	Synthetic Municipal Refuse
Lipids (%)	0.5	-	3.0-9.0	9.7
Crude Fiber (%)	72.0	-	41-61	58.0
Ash (%)	0.5	9.2	4-20	11.0
Volatile Matter (%)	99.5	90.8	96-80	89.0
Crude Fiber/Ash	144	-	2-15	8.1
Bulkweight (pcf @65% Moisture)	5	-	10-15	9
Moisture (%)	5.0	38.6	30-60	47
Average Total Weight in Composters (g-wet)	1000-1400	1850-5000	3000	2000-3000
pH	4.2	7.3	5-8	7.0
Ammonia (mg/g) N as NH_3	0	1.8	0.5-1.4	0.7
Ammonia (mg/g) added	2.0	0	2.0	2.0
Organic Nitrogen (mg/g)	-	8.1	4.8-14	24.3
Total Nitrogen (mg/g)	-	9.9	7-17	27.0
Protein (.625 X Org.N)(%)	-	5.1	3.1-9.3	16.5
Phosphorus Added (mgP/g)	1.3	0	1.3	1.3
Potassium Added (mgP/g)	1.3	0	1.3	1.3

While the solid waste characteristics of these substances are obvious from the table, the following comments on these parameters are necessary.

> The crude fiber analysis has been used as a measure of the paper content in mixed refuse, yet the crude fiber content of newsprint was only 72%.
>
> The average weight used in the bench scale composters was much lower for the newsprint medium because of the lower bulkweight of this material. It was found that the denser the waste material, the greater the amount that could be added to the composter without causing operational difficulties.
>
> The high organic nitrogen content in the synthetic municipal refuse was attributable to a higher meat content.
>
> Nutrient additions did not reflect attempts to optimize but merely to assure adequate levels of necessary constituents for composting without causing toxicity or limiting conditions.

TEMPERATURE

Since maximum stabilization rates of refuse appear to occur under thermophilic conditions and certain experimental gaps exist, a quantitative reexamination of the effect of temperature on composting was in order for high rate systems. This present evaluation emphasized the following points.

> As paper represents a major portion of today's municipal solid waste, the substrate used was deliberately proportioned with a significant amount of paper. Large quantities of a biologically resistant material such as cellulose, the major component of paper, most likely affect the conditions required to maintain an efficient composting operation.
>
> The average rates of degradation of the waste substrates were determined after equilibrium conditions had been established in the composters.

The objective of this phase of the research was to determine the temperature range required for maximum stabilization efficiency of municipal refuse having a high cellulose concentration. Composting experiments were performed at incubation temperature between 22° and 70°C. The materials used as feed included: newsprint; stabilized municipal refuse obtained from the composting plant at Altoona, Pennsylvania; and typical mixed refuse containing 60 to 70% paper.

Initially the bench scale composters were used to produce seed material acclimated at the desired temperature study conditions. Mixtures of a fresh solid waste material and the corresponding seed were then used for batch fed experiments. The composting rates over wide ranges of temperatures were determined. The most promising results, provided by these preliminary studies, indicated the proper temperature range to be further examined in the bench scale composters as continuous systems.

Newsprint

Carbon dioxide production rates presented in a previous progress report have shown the effect of temperature between 30° to 48°C on the degradation of a synthetic refuse containing primarily newsprint. Additional experiments extended the temperature range studied to 62°C. The results are summarized in Table 5.2. The information presented was obtained from experiments using both the bench scale composters and the shake flasks. Seed material used in these studies was acclimated at the desired temperatures, at a pH of 7 and a moisture content of approximately 65%. In the table dm indicates dry mass and VM indicates volatile matter.

TABLE 5.2: COMPOSTING RATES OBTAINED USING NEWSPRINT

Temperature	Mmol CO_2/Day-gdm	Mmol CO_2/Day-gVM
30°C	0.25	0.26
37°C	0.35	0.37
40°C	0.25	0.26
	0.30	0.32
45°C	0.30	0.32
48°C	0.33	0.35
53°C	0.30	0.30
58°C	0.25	0.26
62°C	0.20	0.21

FIGURE 5.5: CARBON DIOXIDE PRODUCTION VERSUS TEMPERATURE FOR CELLULOLYTIC MATERIALS

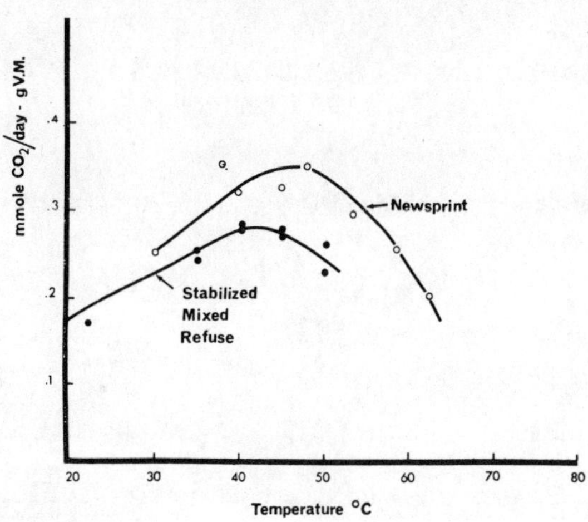

Source: PB 215,722; 1973

The effect of temperature on the rate of composting newsprint is presented in Figure 5.5. The results show that the carbon dioxide production rate increased from 0.26 mmol/day-gVM at 30°C to a maximum rate of 0.35 mmol/day-gVM at 48°C. Decreased biological activity was observed as temperatures were increased to 62°C. This confirms reports in the literature that a predominate number of species of microorganisms capable of utilizing cellulose in nature operate most efficiently within the 45° to 50°C range.

Stabilized Municipal Refuse

A sample of composted municipal refuse was obtained from Altoona, Pennsylvania. Allowing 7 days time for treatment at the plant and transit to the laboratory, the material had been stabilizing for approximately 2 weeks, when used in the composting experiments. Thirty grams of sample were placed in each of ten shake flasks after the moisture content was adjusted to 65%. The pH of the samples was approximately 7.5. Duplicate shake flasks were incubated at temperatures between 22° and 50°C, and carbon dioxide production rates were measured daily for 18 days.

The average composting rates obtained for the mixed refuse are summarized in Table 5.3, and were included in Figure 5.5. The results indicate that the carbon dioxide production rate reached a maximum of 0.28 mmol/day-gVM at 40°C. The magnitude of the stabilization rates obtained were similar to those previously found for newsprint.

This indicates that the composting of the Altoona refuse was essentially complete after 14 days when it was obtained for the experiments. Most of the easily biologically utilized materials were already stabilized. In effect residual cellulose and other resistant materials in the refuse continued to be slowly degraded during the shake flask studies.

TABLE 5.3: COMPOSTING RATES OBTAINED USING STABILIZED MIXED REFUSE

Temperature	Mmol/Day-gdm	Mmol/Day-gVM
22°C	0.155	0.170
22°C	0.240	0.263
35°C	0.220	0.241
35°C	0.230	0.252
40°C	0.25	0.274
40°C	0.255	0.280
45°C	0.245	0.269
45°C	0.25	0.274
50°C	0.205	0.225
50°C	0.235	0.257

Typical Mixed Refuse

Samples containing approximately 50 lb of ground and shredded mixed refuse were collected every other week from an urban area near Philadelphia, Pennsylvania and used as feed during one phase of the composting experiments. Several different batches were used to obtain the information presented. Prior to grinding, glass and metallic components were removed.

The results of chemical analyses, presented in Table 5.4, characterize the batches of mixed refuse used. The variability of the constituents was significant. Crude fiber was found to range from 41.8 to 53.9% and volatile matter varied from 84.5 to 89.3%. The lipid and protein fractions averaged 6.0 to 7.0%, respectively.

TABLE 5.4: CHEMICAL ANALYSIS OF MIXED REFUSE

	Run 6	Run 7	Run 8	Run 9
Moisture	39	60	49	56
Free Air Space (%)	55	35	44	38
pH	5.2	6.0	7.1	6.3
Lipids (%)	6.2	6.9	6.7	4.5
Crude Fiber (%)	53.9	47.7	47.5	41.8
Ash (%)	10.7	12.8	12.1	15.5
Volatile Matter (%)	89.3	87.2	87.9	84.5
Ammonia (mgN/gdm)	.8	.1	.3	.2
Organic Nitrogen (mgN/gdm)	11.9	13.9	8.2	8.0
Protein (.625xN) (%)	9.2	8.7	5.1	5.0
Bulkweight lbs/cuft (moisture 65%)	-	-	12.7	20

Experiments using this mixed refuse were performed in both the shake flasks and the bench scale composters. Temperatures studied extended from 35° to 70°C. For all the experiments, only seed material acclimated to the desired temperature conditions was used. For the typical refuse used, the free air space (Figure 5.8) was close to 30% when the moisture content was 60 to 65%. The pH of the composting mass during the studies was approximately 8.

Carbon dioxide production rates obtained from shake flasks are summarized in Table 5.5 and Figure 5.6 while the results using mixed refuse in the bench scale composters are tabulated in Table 5.6 and Figure 5.7. In both cases, the composting rates increased with temperature reaching maximum values near 60°C. A decrease in the composting rate was observed at higher temperatures.

TABLE 5.5: COMPOSTING RATES USING MIXED REFUSE IN SHAKE FLASKS

	- - - Peak Mmol CO_2/Day- - -	
Temperature	Per gdm	Per gVM
35°C	0.9	1.1
40°C	1.8	2.2
	1.85	2.3
	1.9	2.3
45°C	2.15	2.7
	2.38	2.9
	2.40	3.0
47°C	2.20	2.7
	2.60	3.2
50°C	1.70	2.1
	2.9	3.6
	2.95	3.7
54°C	3.45	4.3
56°C	3.40	4.2
	3.35	4.1
58°C	2.65	3.3
63°C	3.6	4.4
	4.8	5.9
64°C	1.65	2.0
65°C	3.0	3.7
66°C	1.6	2.0

TABLE 5.6: COMPOSTING RATES IN BENCH SCALE COMPOSTERS

	- -Average Mmol O_2/Day- -	
	Per gdm	Per gVM
35°C	0.4	0.5
40°C	1.0	1.25
44°C	1.75	2.35
45°C	1.8	2.3
45°C	2.0	2.7
48°C	2.5	3.35
50°C	1.5	1.88
55°C	2.5	3.5
63°C	2.5	3.7
64°C	3.4	4.6
65°C	2.4	2.9
65°C	1.6	1.95
70°C	2.8	3.5

FIGURE 5.6: CO_2 PRODUCTION RATE VERSUS TEMPERATURE FOR CELLULOLYTIC MATERIALS

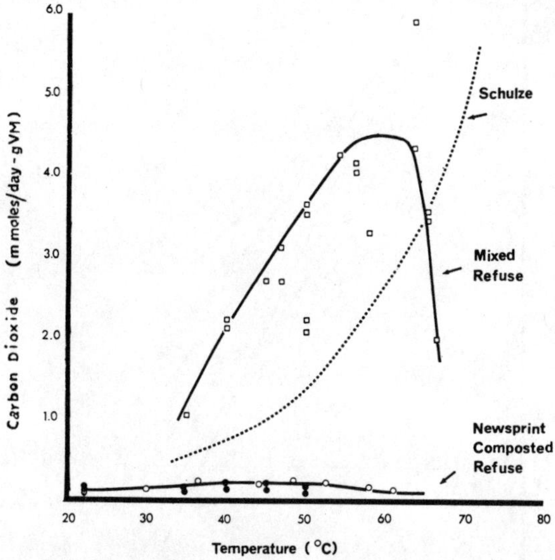

FIGURE 5.7: EFFECT OF TEMPERATURE ON O_2 CONSUMPTION RATE FOR MIXED REFUSE

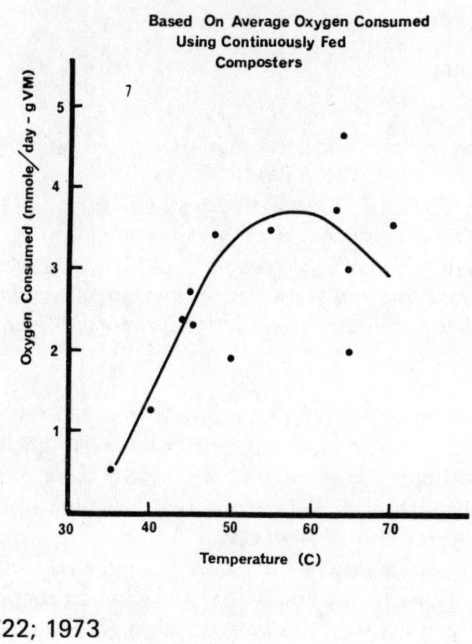

Source: PB 215,722; 1973

FIGURE 5.8: MOISTURE CONTENT VERSUS FREE AIR SPACE FOR ORGANIC SOLID WASTES

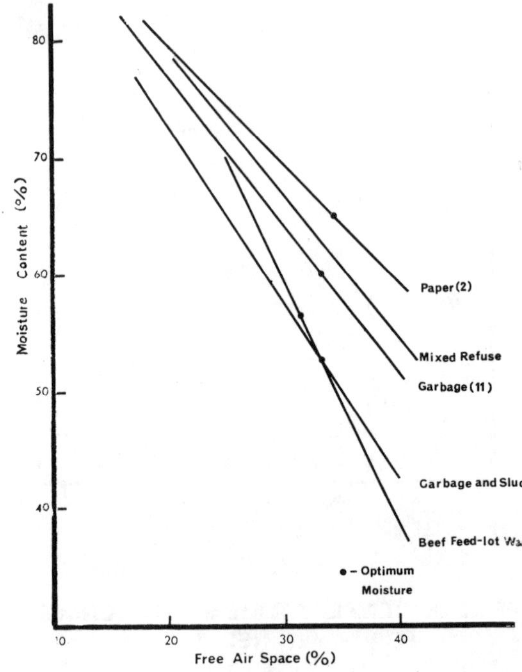

Source: PB 215,722; 1973

The composting rates obtained for the materials previously reported in Figure 5.5 are included in Figure 5.6 along with the results reported by K.L. Schulze in *Aerobic Decomposition of Organic Waste Materials,* NIH (1961). At 60°C the optimum temperature conditions for this investigation, the composting rate obtained for mixed refuse was 61% greater than Schulze's reported value. Degradation rates of newsprint and stabilized municipal refuse were approximately 10% of those found for mixed refuse. Schulze reported maximum results occurred at temperatures close to 70°C.

The difference in the optimum temperature was possibly the results of the following. The composting rates reported by Schulze were not obtained at equilibrium operating conditions. After preliminary results were obtained in batch fed experiments, the bulk data was collected during a single run by measuring instantaneous temperatures and oxygen consumption rates over a 3 day period. These measurements varied significantly during a relatively short period of time. In contrast to Shulze's work the results of the present experiments were obtained when the biological process was operating at close to continuous equilibrium conditions. The carbon dioxide production rates presented in Figure 5.6 represent the peak values obtained in the shake flasks using 50% acclimated seed material. The oxygen consumption rates presented in Figure 5.7 for the bench scale composters represent the average results obtained for 10 to 16 days at constant conditions. The optimum composting rate obtained for mixed refuse under these conditions was between 3.7 and 4.5 mmol/day-gVM.

Schulze's research was limited to refuse containing small amounts of paper while the refuse used in the present study contained 60 to 70% paper and was more typical of solid waste of American cities.

The results (after application of mathematical data fitting techniques) summarized in Table 5.7 indicate temperatures of 40° to 48°C give maximum composting rates for newsprint and stabilized mixed refuse. Maximum composting rates for fresh mixed refuse were obtained at 50°C. 95% of the maximum composting rate was obtained when the temperature was maintained within 5° to 10°C from the optimum value for each of the materials studied.

TABLE 5.7: TEMPERATURE CONDITIONS FOR OPTIMUM COMPOSTING OF CELLULOLYTIC SOLID WASTES

Material	Composting Rate (mmol/day-gVM)	Optimum Temperature
Newsprint	0.35	48°C
Stabilized municipal refuse	0.28	40°C
Mixed refuse	3.7 - 4.5	59°C

Summary

The use of thermophilic temperature conditions for composting organic solid wastes is an established practice. However, only limited information has been available concerning the composting rate of mixed refuse at various temperatures. Further knowledge of the environmental conditions required for improved and more rapid degradation of biologically resistant materials such as cellulose using composting has also been lacking. The composting rates at each temperature obtained in the present investigation are significantly greater than the accelerated rates that have been previously reported in the literature.

More efficient composting rates obtained in this investigation were the result of the following: (a) temperature was controlled within ± 2°C of the desired level; (b) all of the composting rates were obtained with samples containing seed acclimated to within 5°C of the desired study temperature; (c) environmental factors including pH, moisture and free air space were controlled within narrow limits; and (d) aerobic conditions were maintained at all times. This investigation has shown that a high level of environmental control is desirable for improved composting efficiency. The necessity of process design and control in biological waste treatment has long been recognized by engineers and plant operators. The application of similar considerations to solid waste treatment using composting has been lacking and has been partially responsible for many of the past failures reported.

MOISTURE CONTENT AND FREE AIR SPACE

The rate of stabilization of organic material by composting microorganisms is highly dependent upon the environmental factors to which the flora are subjected. Generally, modern mechanical composting processes attempt to create a suitable environment for thermophilic facultative aerobic microorganisms. Criteria for successful composting include: proper temperature and pH, adequate aeration and mixing of organic solids, proper moisture content, suitable structure and particle size and proper concentration of nutrients.

Of the many environmental parameters, moisture content has been referred to as the most important criterion for optimum composting. Sufficient moisture in the waste material is required for maximum efficiency of microbial stabilization. When the moisture content is excessive, the interstices within the organic mass become filled with water and aeration is restricted. As the oxygen concentration is depleted within the material an anaerobic condition develops which results in odors and a significant decrease in the rate of decomposition. (If oxygen were not limiting under such conditions, the biological growth rate would most likely be greater.)

Overall, the suitable moisture contents reported for organic wastes range from 25 to 80%. This wide range of optimum moisture content indicates that a more fundamental and inclusive parameter is needed for understanding the physical and biological interactions controlling composting. It seems fairly obvious that additional factors must affect the optimum stabilization requirements of the various organic waste materials to result in this wide spread in moisture content.

Wiley indicated that the optimum moisture for composting garbage and refuse depends on the composition of the mixture and a summary of information reported is given in Table 5.8. Wiley suggested the concept of effective liquid content as a more valid parameter. The term liquid content includes water and all other free-flowing fluids such as lipids, but not gases, at temperatures reached during thermophilic composting. (J.S. Wiley, *Proceedings ASCE, SED,* No. 1411, 1957)

Using effective liquid content as a parameter, the data for optimum composting of municipal refuse gives better agreement among the various researchers as indicated in Table 5.8. However, this concept alone does not totally explain the significantly different moisture requirements of other organic wastes, including poultry manure, barnyard manure with straw, and night soil. Another factor that affects the moisture requirements of organic waste is particle size. Grinding the material improves the rate of decomposition. However, aeration is adversely affected by finely ground material. For example, the moisture content of a compact organic waste, such as poultry manure, must be low to compensate for the poor aeration characteristics of this type of waste. On the other hand, a higher moisture content is permissible for a porous mixture of straw and manure.

TABLE 5.8: OPTIMUM MOISTURE CONTENT OF HIGH-RATE MECHANICAL COMPOSTING

Location	Material Composted	Moisture Content % Desirable	Moisture Content % Optimum	Effective Liquid Content %
Savannah, Georgia	Municipal Refuse some Paper Removed	55-69	60-65	68.1
East Lansing Michigan	Wrapped Municipal Refuse	55-60	57-58	67.5
Kobe, Japan	Refuse Contains 25% soil	Unknown 60-70 is too high	about 50-55	70.1

A parameter that has been suggested to provide a clearer insight into the mechanisms of composting which is related to the moisture content and the physical structure of the organic waste is called free air space (FAS). The free air space concept has been adapted from soil science and relates the specific gravity (SG), bulk weight (wB) and dry mass (dm = 1 - moisture fraction) of a waste to the air-filled voids within the materials. The working equation to determine the percent free air space for any given waste is: $FAS = 100[1 - (wB \times dm/SG)] \times dm$.

This equation was used to determine the relationship of free air space and moisture content for several organic wastes and the mixed refuse used in the present investigation. The results of these calculations, along with the moisture contents reported optimum for each waste are presented in Figure 5.8. The results show that as paper content increases, the most suitable moisture content for composting an organic waste increases. The moisture requirement for composting increased from 53% for beef feed lot wastes to 65% for wastes containing significant quantities of paper. The corresponding free air space remained constant at 32 to 36%.

Bloodgood, as reported by Schulze, found that over the period 1950 to 1960, the density of mixed refuse in several American cities decreased from 40 to as little as 6 lb/cu ft due to the increase in paper content. This decrease in the bulk weight of refuse has the effect of significantly increasing the free air space.

Therefore, if a uniform free air space requirement for organic wastes exists, as indicated in Figure 5.8, the moisture requirement for composting mixed refuse may be significantly affected by paper content. This phase of the research was undertaken to investigate the relationship of free air space and the rate of composting mixed refuse containing 60 to 70% paper. Oxygen uptake which is related to the composting rate, was measured for samples containing FAS values of 11 to 44% to determine the FAS requirement for maximum aerobic stabilization.

The Warburg Respirometer was used to determine the total oxygen consumption and oxygen consumption rates for each of the samples in Runs 2, 3, and 4. Samples contained, on a dry weight basis, equal portions of fresh refuse and seed. The liquid content of each sample was adjusted with tap water to obtain moisture levels from 90 to 50%.

This corresponds to free air space values of 11 to 44%, respectively, for the refuse used. The lower moisture content was set at 50% since the rate of stabilization of municipal refuse is significantly reduced below this level. The pH of the samples for all runs was approximately 8.

The first run contained samples with free air space values of 23 to 44% (75 to 50% moisture). The free air space range of 11 to 31% (90 to 65% moisture) was studied in the second run. The purpose of overlapping the runs from 23 to 31% free air space (75 to 65% moisture) was to obtain more data in the typical operational range of full-scale composting operations. The final run consisted of samples with a free air space of 11 to 43% (90 to 52% moisture).

The oxygen consumption rates for Runs 2 and 3 are shown graphically in Figure 5.9. The oxygen uptake rate increased as the free air space decreased from 44 to 31% (moisture increase from 50 to 65%). Maximum oxygen consumption rates occurred between 32 and 28% free air space (65 to 70% moisture). This indicates that additional moisture improves the efficiency of the aerobic stabilization. Free air space greater than 28% was sufficient to allow the oxygen requirements of the flora to be met in the refuse used in these studies. As the free air space was decreased from 28 to 11% (70 to 90% moisture), the air filled voids within the waste decreased in volume, thus restricting reaeration. The maximum oxygen consumption was 2.75 and 1.40 mmol/day-gdm, respectively.

The batches of refuse used for Runs 2 and 3 were collected 5 and 10 days respectively before these runs. The low stabilization rates obtained for Run 3 are believed to be the result of degradation of easily utilized components of the refuse during the storage period.

The purpose of the third run was to obtain the oxygen consumption rates for the entire range of free air space between 11 and 43% (90 to 52% moisture). All samples were prepared from a single fresh batch of refuse collected 3 days prior to the run.

The results of Run 4, included in Figure 5.9 show that a maximum oxygen consumption rate of 4.95 mmol/day-gdm was obtained when the free air space was approximately 30% (67% moisture). The oxygen uptake rates obtained with free air space of 11 and 43% were less than 80% of the maximum rate. Balling occurred in the flasks containing samples with less than 23% free air space (75% moisture).

FIGURE 5.9: OXYGEN CONSUMPTION RATE VERSUS MOISTURE AND FREE AIR SPACE

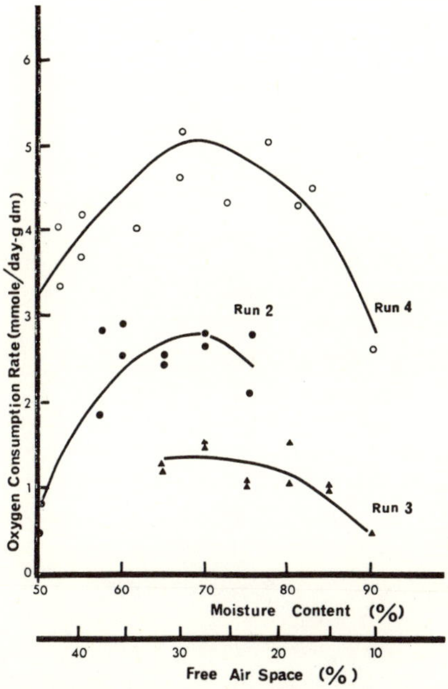

Source: PB 215,722; 1973

The results (after application of mathematical data fitting techniques) summarized in Table 5.9 indicate approximately 67% moisture and 30% free air space were optimum. Further, 95% of the maximum oxygen consumption rate was obtained when the moisture content was between 63 and 79%. The effective liquid content was approximately 70%.

The moisture requirement of 67% for stabilization of refuse containing a significant amount of paper found in this investigation is greater than values reported in past studies using mixed refuse. In previous composting research, the typical refuse treated contained only small amounts of paper. However, the most suitable moisture content for stabilizing municipal refuse biologically may depend, to a large extent, on the amount of paper present in the solid waste.

To determine the magnitude of the effect, calculations were performed using the above equation for mixtures of several solid wastes and paper, maintaining a constant free air space of 32%. The results for mixed refuse, sludge and beef lot wastes show that the moisture requirements increase linearly with paper content with constant free air space.

.BLE 5.9: OPTIMUM CONDITIONS FOR COMPOSTING MUNICIPAL REFUSE

Run	Oxygen Consumption (mmole/day - gdm)	Free Air Space (%)	Moisture Content(%)	Effective Liquid Content(%)
2	2.75	30.0	67.0	70.1
3	1.40	29.5	67.3	69.3
4	4.95	30.0	67.0	70.5

The amount of paper in municipal refuse is not uniform. Wastepaper concentration is generally greater in urban areas than that in rural sections of the country. Furthermore, throughout the country the concentration of paper and other cellulolytic materials in municipal refuse is increasing annually.

The free air space concept suggests a more fundamental and universally applicable parameter for optimum composting of not only municipal refuse, but many other types of organic wastes. Information reported in the literature and shown experimentally indicates that 30 to 36% free air space is required to obtain optimum composting for a wide variety of materials.

RECYCLE

The purpose of this phase of the research was to investigate the effect of recycle on the rate of composting, as measured by the oxygen consumption rate, in completely mixed composting systems using refuse as feed. Preliminary studies were performed using the Warburg respirometer with mixtures of raw refuse and 0 to 90% seed recycle. These results indicated the approximate range of recycle to be used for further experiments in the continuously operated bench scale composters.

Warburg Experiments

Preliminary experiments were performed using the Warburg respirometer to investigate the effect of seed recycle on the composting rate for a typical mixed refuse. Runs 5, 6, 7, and 8 were made with samples of refuse mixed with different amounts of seed acclimated in the bench scale composters operating at the desired temperature conditions to simulate recycle concentrations of 0 to 90%.

Controls containing seed only were included in all runs. Different refuse batches were used for each run with pH varying between 5.4 and 7.6. The pH level of samples at each recycle concentration increased with seed content. The seed had a pH of 7.9 to 8.4. The free air space was approximately 30% in the experiments with a moisture content of 65 to 70%. Studies were performed at temperatures of 50.8° and 59.6°C.

A lag time in obtaining the maximum rate occurred for the samples with less than 50% seed. Oxygen consumption rates typically increased linearly with time at maximum rates for the initial 8 to 10 hours of operation, then gradually decreased after 10 hours. Experimental runs were terminated after 26 to 30 hours. The results show that at 50.8°C, the composting rate was small with no seed and increased to a maximum of 2.9 mmol/day-gVM with 75% seed.

In two of the three runs conducted at 59.6°C, the effect of seed content on the composting rate was similar to the results obtained at 50.8°C. The composting rate was the smallest with zero seed content while the maximum oxygen consumption rate of 3.3 mmol/day-gVM was measured with 75% seed.

The results of the final run were unlike those previously obtained as the oxygen consumption rates were 2.6 mmol/day-gVM with zero seed content and increased to 3.4 mmol/day-gVM with 50% seed. The significant composting rate in the samples with no seed was believed to be the result of the development of thermophilic microorganisms indigenous to the refuse during storage.

Bench Scale Composters

The oxygen consumption rates at seed recycle concentrations of 25 to 75% were investigated using the bench scale composters. Five runs numbered 8, 9, 22, 23, and 24 were made. In the first two runs, batches of an actual mixed refuse were used as the source of feed during the experimental period. The final three studies were made using a mixed refuse prepared in the laboratory.

Each of the composters contained between 2,000 and 3,000 grams of refuse having a moisture content of 60 to 68%. For the mixed refuse used, this moisture content represents from 28 to 36% free air space. Values of free air space with this range were previously shown to provide optimum results for composting. The operating pH level was approximately 8.0 for all composters in the 5 runs. Rotation of the composters, for 2 revolutions each 2 hour cycle, provided mixing of the composting mass.

The composters were adjusted to the desired temperatures between 50° and 70° ± 2°C. Material was removed from each of the composters every 24 to 72 hours, and replaced with fresh refuse to simulate 25, 50 and 75% seed recycle ratios. Recycle concentration refers to the initial percentage of total weight (dry basis) that had been in the composter since the previous feeding and the remaining mass being refuse. On this basis, 100% seed recycle could never be obtained. The mass turnover time in the composters for the five runs was 2.5 or 4 days.

Typically the oxygen consumption rates, measured automatically every 4 hours followed a cyclic pattern between feeding intervals. In composters with 50 and 75% seed, the rate of oxygen consumed increased to a maximum 5 to 10 hours after feeding and gradually decreased until the next addition of refuse. With 25% recycle, the composting rate was more uniform but was less than that of the composters with larger recycles. Representative oxygen consumed data obtained during Run 8, is presented in Figure 5.10. The average composting rates calculated using a planimeter were determined to compare the results for each run. Pertinent results showing the effect of seed on the composting rate recycle are summarized in Table 5.10, and indicate that composting rates were effected by temperature, seed concentration and time lapsed between feeding intervals.

In Runs 8 and 9, using 72 and 75% seed and approximately a four day mass turnover time, an average oxygen utilization rate of 1.9 mmol/day-gVM was obtained at 50°C which increased to 3.7 mmol/day-gVM at 63°C. For 50% seed recycle the average composting rate was 1.9 mmol/day-gVM at 65°C.

FIGURE 5.10: TYPICAL OXYGEN CONSUMED DATA

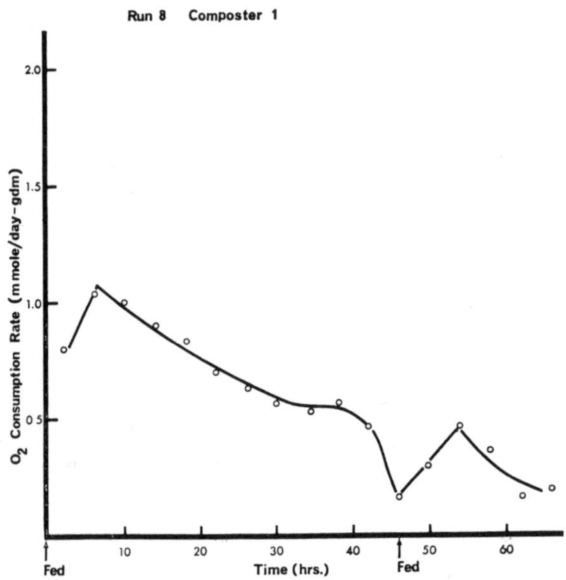

Source: PB 215,722; 1973

The decrease in the composting rates after 10 hours indicates readily degraded components in the refuse were utilized. The feeding interval was reduced from 2 to 1 day to obtain 50% seed recycle and a mass turnover time of 2.5 days. Under these conditions at 64°C, the average value obtained for the oxygen consumption rate was 4.6 mmol/day-gVM. Thus a significant increase in the average composting rate occurred with 50% seed and a decrease in turnover time from 4 to 2.5 days.

TABLE 5.10: EFFECT OF RECYCLE USING BENCH SCALE COMPOSTERS

	Composter Unit No.	Composter Run 8	9	22	23	24
Temperature (C)	1	50	55	51	67	70
	2	65	63	50	62	63
	3	65	64	53	63	63
Mass Turnover Time (Days)	1	4	4	4	4	4
	2	4	4	4	4	4
	3	4	2.5	4	4	4
Seed Recycle Concentration (%)	1	73	72	75	75	75
	2	75	72	50	50	50
	3	50	53	25	25	25
Feeding Interval (Days)	1	1	1	1	1	1
	2	1	1	2	2	2
	3	2	1	3	3	3
Average Oxygen Consumed (mmole/day-gVM)	1	1.9	3.5	2.9	2.6	2.1
	2	2.9	3.7	2.6	1.7	1.3
	3	1.9	4.6	0.3	1.3	0.5
Average Oxygen Consumed (as % of maximum)	1	-	-	100	100	100
	2	-	-	90	65	62
	3	-	-	10	50	24

Unfortunately a mechanical breakdown in the system for obtaining the oxygen consumption data from this composter occurred four days into the run. The composter remained in operation for the remainder of the run, however, visual comparison of the composting mass indicated that the stabilization was less satisfactory using a 2.5 day turnover time. The increased average stabilization rate resulting with the decrease in detention time probably was the result of the larger amount of easily degradable material available in the composter at the increased feeding rate.

The three remaining runs were performed to determine the composting rates using a four day turnover time and seed recycle concentrations of 25, 50, and 75%. Studies were made at temperatures extending from 51° to 70°C. The average composting rates for the various seed concentrations are presented in Figure 5.11 as percentages of the maximum.

At 50° to 53°C, the composting rates with 25 to 50% seed were 10 to 90%, respectively of the the maximum oxygen consumption rates obtained with 75% seed content. At 63° to 70°C, the composting rates with 25 and 50% seed were 25 to 65% of the maximum value obtained using 75% seed.

FIGURE 5.11: EFFECT OF SEED CONCENTRATION ON THE AVERAGE OXYGEN CONSUMPTION RATE

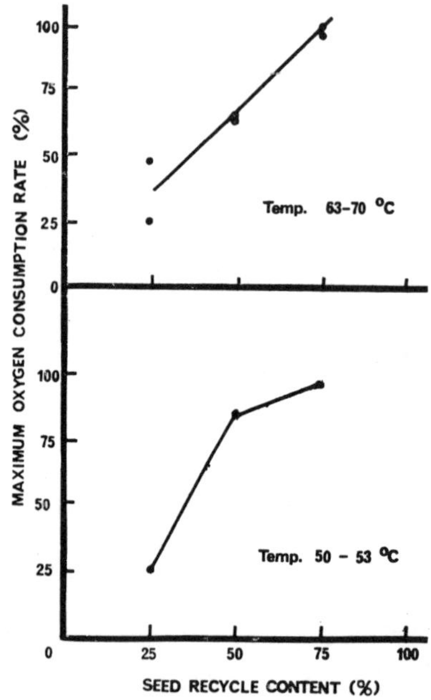

Source: PB 215,722; 1973

At lower composting temperatures (50° to 53°C) less seed recycle was needed to obtain the maximum results on a percentage basis than at 63° to 70°C. The magnitude of the oxygen consumption rates varied for each batch of refuse making further comparison of the results difficult.

In the operation of existing municipal composting plants if large amounts of recycle seed are required for the incoming refuse the rate of composting must be sufficiently accelerated to handle at least the same quantity of refuse treated. Hopefully the accelerated rate would allow a significant reduction in time so that the design treatment capacity would be increased.

On the other hand, refuse disposal costs would be increased if a composting plant had to be expanded to handle the recycle volume. Obviously such expansion would be unwarranted. As shown in laboratory experiments, the bulkweight of refuse increases significantly after composting in laboratory experiments and the volume of seed required must be computed based on weight and would need less composter space. As a typical example, fresh refuse and composted product had bulkweights of 12.7 and 18.1 lb/cu ft respectively yielding a volume reduction of 42%. The moisture content was adjusted to approximately 65% in each case. Mixtures of 25, 50 and 75% seed when mixed with refuse had initial bulkweights of 13.0, 14.5 and 16.0 lb/cu ft, respectively. Calculations were performed using the bulkweight information obtained to determine the increase in bulk volume associated with various amounts of seed recycle. The effect of the increase in bulkweight of the compost on the volume of the seeded refuse was most significant at recycle contents greater than 30%.

Assuming no reduction in time required for composting, the increase in composter capacity needed for each recycle content was determined. If optimum composting is assumed with 75% seed recycle, the greater bulkweight of the seed requires only a 50% greater composter volume.

To compensate for the greater capacity, satisfactory stabilization of refuse would have to be accomplished in significantly lesser time than if no recycle were used. An additional factor should also be considered. Temperatures were maintained in the laboratory composters at desired levels by adding heat electrically. Additional research is required to establish the amount of seed recycle needed to maintain naturally induced thermophilic temperature conditions in well insulated, full scale composting plants. The amount of seed recycle required for optimum composting may also be affected by the following practical considerations.

Temperature – Microorganisms indigenous to refuse are stimulated by warmer temperatures. During the summer months, seed recycle may be decreased as the incoming refuse is more biologically active. When the temperatures are colder, during the winter season, seed recycle may be increased to compensate for the lower biological activity in the incoming refuse.

Storage Time – The development of microbial growth in refuse begins soon after the waste is placed in the sanitary container. Microbial activity may be significant when the refuse is picked up and brought to the composting plant. Depending on the sanitation pickup schedule, refuse may be stored in a trash can for 2, 3, or more days, thus affecting the biological activity.

Shredding – When the refuse is shredded, the composting rate is increased as the surface area for microbial growth is greater, provided that aeration is not restricted. As the seed material is more biologically active, less recycle would be required if the incoming refuse was finely shredded. Shredding also provides mixing, better contact and distribution of the food and microorganisms.

The use of recycle under thermophilic conditions with accompanying decreased treatment times, may result in reducing the overall cost of composting mixed refuse.

However, information is required on a pilot plant scale to ascertain the benefits of recycle. An economic balance could then be made to optimize operating temperature, plant capacity, recycle, and time required for satisfactory treatment.

DEGREE OF STABILIZATION

The composting process relies on the biological degradation of organic material present in mixed refuse to produce a relatively stable humus, which can be safely returned for use in the ecological cycle. The nature of mixed refuse is such that a wide variance exists, regarding the composition of the solid waste.

In general, the organic fraction contains approximately 70% paper and 30% food wastes. The nonhomogeneous nature of the refuse makes measurement of the degree of stabilization difficult. It was desirable to use a scientific test procedure to determine composting efficiency which could operate independently of the characteristics of refuse. Chemical oxygen demand (COD) was used for this purpose.

The objective of this phase of the research was to measure the amounts of soluble materials leached from refuse and composted product. Procedures for the water, base and acid soluble extractions were detailed under Specific Analyses page 108. A rapid procedure for measuring COD was used to measure the leachate concentration.

Preliminary studies on the COD of leachate of the refuse and composted product were made during Composter Runs 15 and 16 and reported in Table 5.11. Based on the total soluble material extracted from the samples, the composted products had lower COD leachate concentrations than the refuse. The reductions ranged from a minimum of 8.2 to a maximum of 61.9%.

TABLE 5.11: REMOVAL OF COD FROM LEACHATE BY COMPOSTING

Composter Run	Sample	Soluble COD (mg/gdm) Water	Base	Acid	Total	COD Removal(%)
15	Refuse	140	197	26	363	-
	Composter 1	169	83	23	275	24.3
	Composter 2	210	93	29	333	8.2
	Composter 3	136	86	45	267	26.5
16	Refuse	64	146	10	220	-
	Composter 1	31	48	5	84	61.9
	Composter 2	46	74	8	128	41.8
	Composter 3	23	58	15	96	56.4

Wide variation of the COD removals resulted since each composter was operating with different mass turnover times, seed concentrations, pH and temperature conditions. Also the refuse used in Run 16 was of poor quality. Most significantly, however the results show that the soluble matter in the compost was lower than that in the refuse. This indicated that the technique of measuring the soluble matter in solid waste had potential as a meaningful parameter in determining the degree of stabilization in composting.

The rate of COD removed in the leachate with time was determined in Run 19, Composter 2. The refuse used as feed was prepared in the laboratory and stored at 4°C until used. The operating conditions of the composter at the time of the study were: temperature 58°C, pH 8.6, moisture 60%, and seed concentration 50%. The initial mass turnover time was approximately 2 days. By suspending feeding during the experiment the turnover time was extended to 9 days.

The rate of COD removal for duplicate samples was determined. After a 4 to 5 day retention period, the maximum removal of soluble materials has occurred. For the water-soluble fraction, the maximum percent COD reduction was approximately 68%. For the base soluble fraction the maximum was approximately 30%. The concentration of acid soluble materials was not determined for these samples. The results of the study indicated that in general a 4 to 5 day mass turnover time was satisfactory for maximum removal of soluble materials present in mixed refuse under the given experimental conditions.

The removal of soluble materials as judged by leachate COD, suggests a rational procedure for determining the degree of stabilization obtained using the composting process. Minimum leaching from composted solid wastes is essential if the treated waste is to be used for aesthetically and hygienically safe land reclamation or other uses.

EFFECT OF pH

The objective of this phase of the research was to investigate the effect of pH on the composting rate of mixed refuse. Experiments were performed at temperatures controlled near 46° and 60°C, using the bench scale composters and the Warburg respirometer. The composting rate was estimated by oxygen consumption and carbon dioxide production.

Microscopic examinations of the composting mass were made to correlate the microbial flora present with the various experimental conditions. Knowledge of the interrelationship of pH, temperature and microbial population, to the stabilization rate of mixed refuse is desirable for an optimized composting process. Preliminary experiments were run using the Warburg respirometer.

Bench Scale Composters

The results of the batch composting experiments using the Warburg respirometer at 59.6°C, indicated that pH values near 8 were optimum for composting as determined by oxygen consumption and carbon dioxide production rates. Acclimation of composting mass to the lower pH levels used in the study did not provide a significant improvement in the composting rates.

Information obtained on the effect of pH, using the bench scale composters is summarized in Table 5.12. Each of the average composting rates given represent 10 to 16 days of continuous operation at the desired conditions. The composters were controlled at temperatures of 46° to 59°C.

TABLE 5.12: EFFECT OF pH IN BENCH SCALE COMPOSTERS

	Composter	pH	Temp. °C	Average Oxygen Consumption mmole/day-gVM	Average Carbon Dioxide Production mmole/day-gVM
Composter Runs 19 - 21	1	6.5	59	1.87	1.87
	1	6.9	58	2.16	2.05
	1	6.0	46	3.82	4.39
	2	8.3	58	8.20	8.10
	2	8.0	58	5.45	4.35
	2	7.8	57	6.31	5.25
	3	7.5	57	3.85	3.85
	3	6.5	57	3.1	2.58
	3	7.0	57	3.7	4.20
Composter Runs 12 - 15	1	7.0	54	-	-
	1	7.0	54	3.89	3.06
	1	6.8	54	1.72	1.67
	1	7.6	55	1.88	1.88
	1	6.9	55	3.16	3.93
	1	6.9	57	3.52	2.93
	2	7.9	55	-	-
	2	7.5	55	3.0	2.78
	2	7.5	56	-	1.11
	2	7.6	56	-	2.15
	2	8.1	58	7.3	7.94
	2	7.8	56	6.63	6.07
	3	8.0	55	-	-
	3	8.0	58	8.34	5.55
	3	7.5	57	2.84	2.84
	3	7.8	56	1.94	2.15
	3	7.6	56	6..27	6.76
	3	6.9	56	3.64	5.36

The pH ranged from 6.0 to 8.3. At temperatures of 54° to 59°C the composting rates were minimum at acidic pH levels, and increased to maximum values of 8.2 mmol/day-gVM under alkaline conditions. In one composter, which was operating at pH 6.0, the temperature was decreased to 46°C. After 14 days of operation, the average oxygen consumption rate was 3.82 mmol/day-gVM. This is a significant increase in the composting rate when compared to the rates obtained at 54° to 59°C with similar pH levels.

However, the rate of composting at pH 6 and 46°C was less than 50% of that obtained at pH 8 and 54° to 59°C. The flora present at pH 6 and 46°C was found to contain a significant amount of filamentous growth, which was assumed to be heat resistant fungi. No filamentous growth was observed at 54° to 59°C. Fungi were better suited to the acidic pH level at 46°C than at the higher temperatures. Overall, composting proceeded most efficiently at thermophilic temperatures when the pH was about 8.

Fluctuation Using Mixed Refuse

The composters were fed at 24, 48, or 72 hour intervals to obtain the desired seed recycle concentration and mass turnover times. The daily pH measurement of the composting mass was determined at least 18 to 24 hours after fresh refuse was added. Typically the pH was 5 to 6 for the refuse and 7.5 to 8.5 for the composting mass unless chemical additions had been made to provide pH control.

In many of the composter runs using mixed refuse, the oxygen consumption rates varied significantly between feeding intervals. The composting rate began to increase soon after feeding and reached a maximum 6 to 10 hours later and decreased thereafter until more refuse was added. An experiment was performed to study the fluctuation of pH during feeding intervals and its possible influence on the composting rate.

Measurement of the buffering capacity of composted mixed refuse was determined using a sample of compost having an initial pH of 8.2. The titration with standard sulfuric acid results show that the compost used for the experiment had very little buffering capacity. Small amounts of acidic or basic constituents in the refuse could have resulted in a significant change in the pH level during composting. The volatile acid and ammonia concentrations in refuse was measured and found to be approximately 5 and 0.3% respectively. The volatile acids probably account for the characteristically acidic pH of refuse.

Composter 2, Run 9 was used in this pH experiment. The operating conditions at the time of the pH study were: temperature 63°C, moisture 66%, seed recycle concentration 72%, and mass turnover time 3.6 to 4.1 days. The pH determinations were made by removing samples from the composter immediately after feeding and at frequent intervals thereafter.

The initial pH level was 5.8 and increased to 7.9 after four hours, and to 8.2 after 24 hours. The increase in the pH level during composting resulted because of the low buffering capacity of the refuse used. The reduction of volatile acids, combined with presence of ammonia released from protein wastes, enabled alkaline pH conditions to be reached after 4 hours. The maximum composting rate, obtained 6 to 10 hours after feeding, occurred at alkaline pH levels.

The average pH value reported for each of the composters in Table 5.12 was representative of the operating conditions of the systems for at least 20 hours a day during the runs. The initial four hour fluctuation of pH probably did not have a significant effect on the average composting rate. Obvious economic advantages result as maximum stabilization rates using composting occur under naturally induced conditions at the slightly alkaline pH of about 8.

NUTRIENT REQUIREMENTS

Nutrients, primarily nitrogen and phosphorus, are known to be important in biological waste treatment. To quantize the nutrient requirements of mixed refuse in composting, Warburg runs were made at 59.6°C using different nitrogen and phosphorus levels. The objective of the experiments was to define the critical concentrations of nutrients necessary to insure that maximum biological activity was maintained. It was also desired that the amounts of chemicals added be minimal in order to maintain an economical process.

Nitrogen

In the nitrogen study Warburg Run 11 was set up using a number of duplicate flasks, each containing mixtures of 75% seed and 25% refuse and ammonia concentrations ranging from 2.6 to 8.3 mg NH_3-N/gdm. The remaining flasks contained an ammonia concentration of 21 mg NH_3-N/gdm with seed recycles of 25 and 50%.

The seed material employed for this run was taken from a bench scale composter which had received 2.0 mg/NH_3-N/gdm refuse with each feeding. The ammonia in the refuse was 0.6 mg NH_3-N/gdm. While control samples were used in the Warburg study in which no ammonia was added, there was still a significant quantity of nitrogen available to the microorganism from the seed and biological degradation of proteinaceous materials. The moisture content was approximately 65% in all samples and the pH was close to 8.

The composting rates did not change with nitrogen concentration for samples with 75% recycle. The concentrations examined (2.6 to 8.3 mg N/gdm) may lie somewhere in the optimum range, since there was no significant increase or decrease in oxygen uptake rate with the concentration differential. In the bench scale composters, ammonium chloride added with each feeding of refuse provided the equivalent of 2.0 mg NH_3-N/gdm, and residual ammonia was found in the composted products. Ammonia concentrations of 0.5 to 1.4 mg N/gdm were measured in the mixed refuse as received. Therefore, sufficient nitrogen was available in all the bench scale studies used to obtain optimum composting rates.

For samples with 25 to 50% recycle, reduced oxygen consumption rates were obtained with ammonia concentrations of 21 mg NH_3-N/gdm. The decreased rates were probably caused by insufficient seed recycle under the experimental conditions used.

In bench scale composters operating with four day mass turnover times and 25 to 50% recycle, (Runs 22, 23, 24), detectable ammonia odors occurred after 24 hours using 2 mg NH_3-N/gdm with each feeding. In these composter studies the operating conditions had been adjusted to reduce the recycle to 25 and 50%, and maintain the desired turnover times, by feeding refuse once every 2 or 3 days.

Volatilization of ammonia was attributed to losses by chemical means, as the temperature was close to 60°C and the pH was alkaline. The results indicate that both the concentration and quantity of ammonia added as a nutrient is critical for optimum composting. Although the minimum ammonia required for optimal microbial activity was not defined, it was determined that excessive quantities were partially lost before utilization.

Phosphorus

In Warburg Run 12 performed at 59.6°C to determine the effect of phosphorus sixteen flasks were set up, each containing refuse seeded with 50% recycle. To each of the samples, different concentrations of potassium phosphate ranging from 0.6 to 15.0 mg P/gdm were added. The composting rates decreased significantly with increasing phosphate concentration. Unfortunately the concentration of phosphate was not the only variable in this experiment.

The pH varied initially from 6.2 to 8.2 but tended to increase during the period the oxygen consumption rates were determined. More care should have been taken to control pH. The results are inconclusive since it cannot be determined whether the decrease in composting rates was due to the phosphate concentration or pH.

During the research, a concentration of phosphorus equivalent to 1.3 mg P/gdm was added at each feeding interval to the bench scale composters to provide for the nutrient requirements. In biological waste treatment processes, the ratio of nitrogen to phosphorus required is satisfactory at an N to P ratio of 5 to 1. Considering the concentration of nitrogen that was found suitable in the previous discussion, phosphorus additions were in excess of the minimum requirements in all bench scale composting experiments conducted.

REFUSE STORAGE CONDITIONS

The activity of microorganisms indigenous to mixed refuse can be significant. Both ambient temperature and storage times would be variables which can effect the initial degradation rate. The effect of these conditions on the composting rate was demonstrated as the research project progressed.

In Composter Runs 16, 17 and 18 the same batch of mixed refuse was used as feed. This refuse had been stored in several metal containers at room temperature for up to 7 weeks. A decrease in the composting rate with time was noted. The decreasing trend of the composting rates occurred in composters operating at pH levels of 6.7, 7.5 and 8.2. The results indicated utilization of the easily degradable materials during storage. Normally, refuse was stored for less than 2 weeks at 20°C and deterioration was less significant. In an attempt to further minimize the deterioration of the refuse, batches of the waste material was stored at 4°C.

Both of these methods of storage are not typical of refuse collection procedures which are such that the solid waste is not normally stored in a sanitary container more than 3 to 7 days before deposition at a disposal site. Temperatures approaching 4°C would only be characteristic of severe winter conditions, but would be critical as far as designing composter plants. In order to gain knowledge of the effect of storage conditions on the composting rate of mixed refuse, Warburg Run 13 was made. For the Warburg run, samples were prepared from material which was initially stored at 4°C and then incubated at each temperature and storage time indicated in Table 5.13. Samples stored at 4°C were included. Seed recycle was varied from 15 to 75%. The results of the experiment are also summarized in Table 5.13 according to storage conditions. Lag times in obtaining the maximum oxygen consumption rates ranged from 2 to 5 hours for samples with 50% recycle or less. With 75% seed recycle, when refuse was stored at 4°C, the maximum rate of oxygen consumed was obtained with no lag.

TABLE 5.13: EFFECT OF STORAGE CONDITIONS – WARBURG RUN 13

Flask	Storage	Composting Rate ($mg\ O_2$/hr-gdm) Initial Uptake	Maximum Uptake	Lag Time (hrs)	% Seed
1	4°C	4.30	4.3	None	75
2	4°C	4.30	4.3	None	75
3	4°C	3.12	-	Failed*	50
4	4°C	3.12	5.90	2	50
5	20°C	3.12	5.27	2.5	50
6	1 day	3.12	4.57	4	50
7	20°C	2.63	4.65	3.5	25
8	1 day	2.63	5.14	2.5	25
9	Room Temp	3.08	4.85	4	50
10	1 day	3.08	-	Failed*	50
11	Room Temp	2.62	2.62	No lag	25
12	1 day	2.62	-	Failed*	25
13	Room Temp	2.94	4.88	3	25
14	1 day	2.94	4.25	5	25
15	Room Temp	3.02	5.74	3	15

* Leak developed during run.

Maximum composting rates with as little as 15% seed recycle were obtained after a 3 hour lag period. The solid waste for these samples had been stored for 2 days at room temperature following storage at 4°C. Storage conditions have been shown to affect the composting rate. Larger recycle concentrations were required when the refuse was stored at lower temperatures. Under warmer storage conditions, less recycle was needed, as microbial activity in the refuse was more significant. The seed requirement for optimum composting may vary with both the geographical location of the plant and season of the year. It is evident from the results of this study that full scale composting operations should be designed to allow for flexibility in recycle capabilities.

PAPER CONTENT

When considering the effect of temperature it was found that the composting rate of newsprint as measured by oxygen consumed and carbon dioxide produced was 0.2 to 0.3 mmol/day-gVM. The composting rate of mixed refuse containing 60 to 70% paper was as high as 4.0 to 5.0 mmol/day-gVM, an order of magnitude above the previous result. The typical municipal refuse of American cities contains large amounts of paper and other cellulolytic materials and the concentration of paper is generally increasing.

Due to its significant concentration the effect of high paper content on the composting rate of mixed refuse was investigated in Warburg Run 14. Seven batches of mixed refuse were prepared in the laboratory with paper content varied from 50 to 100% and the remaining portion composed of vegetable and meat wastes in a 2:1 ratio.

For the Warburg run 16 samples were prepared, 14 containing a mixture of 50% seed and 50% of the prepared refuse. Two of the flasks were controls and contained 100% acclimated seed. The summary of the contents of each flask is presented in Table 5.14. Calculations were made to determine the total percentage of paper in each flask. It was assumed that the seed material contained approximately 70% paper and the paper concentration in the Warburg flasks ranged from 60 to 85%. The moisture content of all samples was 65 to 68%; the pH was close to 8.0 and the temperature was 59.6°C.

TABLE 5.14: MIXED REFUSE PREPARED FOR WARBURG RUN 14

Flask	% Seed	% Paper	% Vegetable Scraps	% Meat Scraps
1&2	50	25	16.5	8.5
3&4	50	25	16.5	8.5
5&6	50	30	13.5	6.5
7&8	50	35	10.0	5.0
9&10	50	40	6.6	3.4
11&12	50	45	3.4	1.6
13&14	50	50	0	0
15&16	100	0	0	0

The oxygen consumption rates at each paper concentration is shown in Figure 5.12. The graph shows a significant decrease in the composting rate when the paper content is greater than 75% thus indicating the resistant nature of cellulose, the primary component of paper. The constant composting rate for samples with less than 75% paper indicates stabilization of the readily degraded fractions of vegetables and meat during the initial 8 to 10 hours of the Warburg run.

FIGURE 5.12: EFFECT OF PAPER CONTENT ON OXYGEN CONSUMPTION RATE

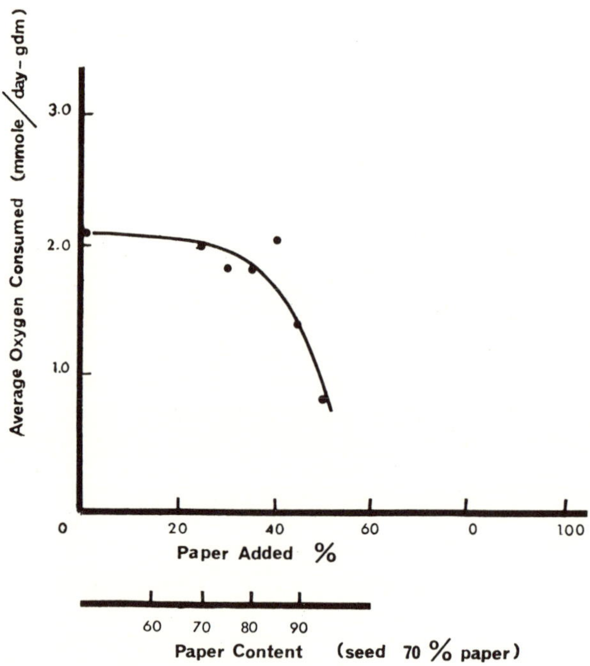

Source: PB 215,722; 1973

Future studies on the effect of paper should consider temperature since Warburg Run 14 was performed at 59.6°C, which is close to the optimum temperature for composting refuse. Probably the temperature should be lower, perhaps 50°C, to optimize the degradation of cellulose. In practice, an optimized composting process might employ a short term high temperature stage for maximum stabilization of readily utilized components in the refuse followed by a longer term second stage controlled at lower temperatures for improved cellulose decomposition. In the bench scale composting experiments, operating with a 4 day mass turnover time, the composted refuse created no objectionable odor problem and appeared to contain mostly paper. Thus, another potential use of composting would be as a pretreatment step for reclaiming the paper in municipal refuse. Once the easily utilized components are removed by biological action the concentrated paper fraction of the refuse can be recycled into the economy as scrap paper for reprocessing.

Reclaimed paper could serve as the prime raw material of many products including fertilizer, clean landfill, building materials as well as paper products. It certainly does not appear possible to fully degrade paper using controlled environmental conditions in composting in 4 days or less.

SUMMARY

Optimum stabilization levels were reached in the laboratory using a high degree of experimental control in continuous composting systems. In conclusion, the environmental conditions recommended to obtain the maximum composting rate of typical municipal refuse and other solid wastes are summarized in Table 5.15.

TABLE 5.15: ENVIRONMENTAL CONDITIONS FOR OPTIMUM COMPOSTING

Substrate	Parameter Controlled	Optimum Value
Newsprint	Temperature	48C
Stabilized mixed refuse	Temperature	40C
Mixed Refuse	Temperature	59±5C
	Moisture	67±3%
	Free air space	32±3%
	Seed recycle	50 to 75%
	Mass turnover Time	4 to 5 days
	pH	8.0±0.5
	Nitrogen added (as NH_3-N)	2.0 mg/gdm (0.2%)
	Phosphorus added (as P)	1.34 mg/gdm (0.134%)
	Air requirements (cu ft/day-lbVM)	16.7-20.3
	Air residual (%)	10

The average reductions of the components in the refuse after composting are summarized in Table 5.16. The results represent 24 runs, the data of which was obtained after 10 to 30 days of continuous operation under the desired conditions. The mass turnover time was normally 4 to 5 days.

TABLE 5.16: AVERAGE REDUCTION OF MIXED REFUSE BY COMPOSTING

Refuse Component	Reduction (%)
Total Matter	40
Volatile Matter	45
"Other" Volatile Matter (Carbohydrates)	65
Lipids	86
Crude Fiber	30
Protein	22
Bulk Volume*	42-50

* Mass adjusted to 65% moisture

A short summary of the major conclusions follows.

Temperature: In the laboratory experiments temperature has been shown to have a significant effect on the composting rate of solid wastes containing large amounts of paper. For freshly collected municipal refuse, containing 60 to 70% paper, an optimum composting rate of 3.7 to 4.5 mmol CO_2 and O_2/day-gVM was measured at 59 ± 5°C. The composting rate of paper was less than 10% of the rate obtained for mixed refuse.

Moisture Content and Free Air Space: The optimum moisture content and free air space for composting typical mixed refuse were 67 and 30% respectively. Calculations performed using information presented in the literature indicates similar free air space results for many other organic wastes having significantly different moisture requirements. The free air space concept suggests a more fundamental and universally applicable parameter for composting not only municipal refuse, but many other types of organic solid wastes.

Seed Recycle: Under the conditions studied in the laboratory, 50 to 75% seed recycle was required to obtain maximum results in continuous composting experiments. The amount of seed recycle required for optimum thermophilic composting is affected by the condition of the incoming refuse. Most significantly, ambient temperature, storage time in sanitary containers before collection and the degree of size reduction or shredding of refuse influence the amount of recycle required in composting.

Mass Turnover Time and Degree of Stabilization: The reduction of COD of leachate of mixed refuse was used to indicate the degree of stabilization obtained using the composting process. With mass turnover times of 4 to 5 days, the reduction of the water and base soluble components in refuse reached maximum levels of 68 and 28% respectively. As a test for stabilization the COD of the leachate offers some promise.

pH: Laboratory investigations, using controlled conditions, indicated that maximum thermophilic composting rates were obtained at pH 7.5 to 8.5. Stabilization efficiency was significantly decreased at lower or higher pH levels.

Nutrients: The mixed refuse used as feed for the composting studies was supplemented with nitrogen and phosphorus salts to provide these necessary nutrients. Maximum composting rates were obtained using nutrient concentrations of 2.0 mg NH_3-N and 1.34 mg P, added for each gram of refuse fed (dry basis). Under certain experimental conditions, detectable ammonia odors occurred during composting indicating excess nitrogen was present. The minimum nitrogen and phosphorus requirements for optimum composting were not defined.

Storage Conditions: For many composter runs, refuse was stored in metal containers at 4°C or room temperature until needed for the experiments. This method of storage may have altered the stabilization characteristics of the mixed refuse feed. Composting rates decreased with the length of refuse storage at 20°C. Larger recycle concentrations were required for maximum results when the refuse was stored at the lower temperatures as the natural seeding effect was diminished.

Paper Content: The effect of increasingly higher concentrations of paper on the composting rate of mixed refuse was investigated. A significant decrease in the composting rate occurred in refuse samples having greater than 75 to 80% paper.

Compost as Soil Amendment

The objectives of the studies, described below, by C.C. Hortenstine and D.F. Rothwell in PB 222,422 were (1) to evaluate composted municipal refuse as a source of plant nutrients, (2) to determine the effects of adding composted municipal refuse to the soil on water-holding capacity, cation exchange capacity, pH, conductivity, extractable plant nutrients, etc., (3) to determine if phytotoxic substances were present in composted municipal refuse, and (4) to determine the effects of adding composted municipal refuse to the soil on microbial activity, in particular, on nitrification and carbon dioxide evolution.

MATERIALS AND EXPERIMENTAL METHODS

Largo Compost

Greenhouse – Composted municipal refuse used in this experiment was obtained from an industrial composting plant which was under contract to compost the refuse from Largo, Florida. Chemical analyses of the compost are shown in Table 6.1. As obtained from the plant, the compost was 44% H_2O on a wet weight basis (79% H_2O dry weight basis). The compost was air-dried and screened through hardware cloth with 6 mm openings prior to use.

In order to determine the effects of compost on germination, varying amounts (0, 10, 20, 40, 80, 160, and 320 grams) of compost were shaken in 500 ml of distilled H_2O for ½ hour and filtered. The extracts were stored at 7°C until used. At that time, filter paper discs set in petri dishes were soaked with 5 ml of the extract and the dishes were arranged in randomized blocks of four replications. Twenty-five seeds of oats, *Avena sativa* L. (Suregrain); turnip, *Brassica rapa* L. (Purple Top White Globe) and pearl millet *Pennisetum typhoideum* L. (Rick., Gahi) were planted on filter paper. After 5 days at 21°C, each group of seedlings was removed, counted, and measured from plumule tip to radicle tip. The filter paper and extracts were renewed between each species planted.

Leon fine sand was air-dried, screened, and limed at the rate of 4 metric tons per hectare. Soil and garbage compost were thoroughly mixed in a twin-shell blender and the mixtures were put into plastic pots (2,600 cu cm volume) and arranged in randomized blocks with four replications. Oats, turnips, radishes, and pearl millet were planted consecutively, thinned to a uniform stand, and water was replenished daily. Oat foliage was cut at 5 weeks, turnip and radish at 6 weeks, and millet at 10 weeks. Between each crop, the soil was screened to remove roots and repotted.

The Leon soil series belongs to the Spodosols Order, more commonly referred to as Ground-Water Podzols. The most distinguishing feature of Leon soils is a prominent spodic horizon (organic pan) within a depth of 76 cm. The Spodosols comprise 3,737,443 hectares (9,235,223 acres), or 26.60% of Florida's land area. Under proper water and fertility management, these soils are used profitably for vegetable, citrus, and livestock agriculture. The Leon fine sand used in this experiment had a pH of 4.4 and contained 85 ppm Ca, 30 ppm Mg, 13 ppm K, and 2 ppm P.

Plant tissue was dried at 70°C, weighed, and ashed at 500°C for chemical analyses. Nitrogen was determined by the micro Kjeldahl method, P was determined colorimetrically, and K was determined by flame photometry. Soil was sampled after the millet plants were harvested, air-dried, and extracted with NH_4OAc (pH 4.8) for Ca, Mg, P, and K determinations.

Cation exchange capacity (CEC) was determined by the NH_4OAc method, soil pH was determined in a 1:1 soil: water mixture, and total soluble salts (TSS-mmhos x 640) were determined by electrical conductance in a saturated soil extract. Water-holding capacity was determined by the pressure membrane technique described by L.A. Richards, *Soil Science,* 68:95-112, July-December 1949.

Field – This experiment was conducted in 17.6-liter crocks (hereafter referred to as pots) set in the soil. The bottom of each pot was covered with a 2-inch layer of gravel which was, in turn, covered with a layer of glass-wool upon which the treated soil used in the experiment was placed. A glass-wool plug inserted in the drainhole and in contact with the surrounding soil allowed free drainage through the pots.

Individual treatments of garbage compost, sewage sludge, cow and chicken manures, and 10-10-10 fertilizer were mixed in a motor-driven cement mixer with 13.7 kg of Leon fine sand which was limed at the rate of 4,500 kg/ha. Treatments were arranged in the pots in randomized blocks with four replications. Fifty oat seeds, *Avena sativa* (Suregrain), were planted in each pot and two cuttings of oat foliage were harvested. After the oat roots were removed, radish *Raphanus sativus* (Early Scarlet Globe) seeds were planted and the seedlings were thinned to 24 per pot when 10 days old.

The organic materials used in this experiment were commercial products except for the garbage compost which was obtained from a composting plant in Largo, Florida. All materials were air-dried and ground to pass a 20-mesh screen before incorporation into the soil. Laboratory analyses of organic materials are presented in Table 6.1. The high salt content of all materials shown would indicate a possible danger to salt-sensitive plants if excessive amounts of these materials were applied to the soil, especially where rainfall was not sufficient to carry the salts below the root zone.

Plant material was dried at 70°C, weighed, and ashed at 500°C for chemical analyses. Nitrogen was determined by the micro Kjeldahl method, P and B were determined colorimetrically, K was determined by flame photometry and the other plant nutrients (Ca, Mg, Cu, Mn, and Zn) were determined by atomic absorption.

After the radishes were harvested, soil samples were removed from each pot, air-dried, and screened through a 0.5 mm sieve for chemical analysis. Soil pH was determined in a 1:1 soil:water mixture, total soluble salts were determined by electrical conductance in a saturated soil extract, cation exchange capacity (CEC) was determined by the NH_4OAc method, and soil Ca, P, K, and Mg were determined in an NH_4OAc (pH 4.8) soil extract. Soil water-holding capacity was determined as moisture equivalent (ME) which is the ability of a soil to hold water under a centrifugal force 1,000 times that of gravity.

Microbiological – The surface (0-to-15-cm) layer of an Arredondo fine sand was used for all decomposition studies. The soil had been planted to agronomic crops for many years and had pH 6.4. Nitrification had been previously demonstrated to occur in this soil. The Arredondo soil had not been classified in the comprehensive system, but will probably be included with the Quartzipsamments.

Organic materials used in these experiments were garbage compost, chicken manure, cow manure, and sewage sludge. All materials except the garbage compost were obtained from local commercial sources. Garbage compost was obtained from a composting plant in Largo, Florida. These organic materials (Table 6.1) were dried at 70°C and ground in a Wiley mill to pass a 20-mesh screen. Each material was mixed with soil at specified ratios and moisture adjusted to 100 millibars of water suction.

The amended soil was then incubated at 28°C. Evolved CO_2 was absorbed in standard NaOH solution and precipitated as $BaCO_3$ by addition of sufficient $BaCl_2$ solution to insure complete precipitation. Excess NaOH was titrated with standard HCl. Carbon dioxide evolution was determined throughout the various studies. The amounts obtained were added to the previous evolution and reported as totals. Phenolphthalein was used as the indicator. Nitrate-nitrogen was determined by the phenoldisulfonic acid method.

TABLE 6.1: CHEMICAL CONSTITUENTS OF ORGANIC MATERIALS

Material	pH	ash	C	TSS	N	P	K	Ca	Mg	Mn	B	Cu	Zn
		%								ppm			
Cura Compost	7.5	16	33	1.73	0.76	0.17	0.30	0.99	0.07	106	-	29	196
Fairfield Compost	5.7	-	39	0.02	2.27	0.45	0.20	0.01	-	8	16	-	294
Largo Compost	6.9	44	31	2.30	1.20	0.26	0.38	1.30	0.07	130	25	125	250
Gainesville Compost	6.9	42	32	2.15	0.58	0.23	0.21	2.14	0.12	140	40	-	540
Milorganite Compost	4.1	27	-	2.20	5.46	0.28	0.34	0.30	0.16	23	5	90	180
Cow Manure	6.7	50	-	4.40	1.32	0.52	2.52	1.29	0.27	81	6	16	38
Chicken Manure	7.2	33	-	4.05	2.64	0.48	1.80	1.85	0.38	133	17	32	196

The influence of garbage compost on numbers of soil microflora was investigated by mixing 0, 1, 5, and 10 grams of compost with 100 grams of soil and incubating for 26 days at 28°C. Plate counts, in quintuplicate, were made periodically for relative numbers of fungi on rose bengal agar and for bacteria on soil extract agar.

Garbage compost, cow manure, chicken manure, and sewage sludge were used as amendments to Arredondo fine sand at 0, 2, 4, 6, 8, and 10 grams per 100 grams of soil, replicated three times. The amended soil was incubated for 65 days. Carbon dioxide evolution was determined throughout the study and NO_3-N was determined at the end of 65 days.

Ground garbage compost was also used at 0, 10, 20, 30, 40, 50, 60, 70, 80, 90, and 100 grams per 100 grams of soil and incubated for 101 days. Carbon dioxide evolution was determined at various intervals during the investigation.

Another study was conducted to investigate the influence of mixing garbage compost with other organic materials. In this study, garbage compost was mixed with cow manure, chicken manure, and sewage sludge in equal amounts by weight and added to the soil at levels of 0, 1, 2, 3, 4, and 5 grams of each material per 100 grams of soil and incubated for 64 days. Carbon dioxide evolution was determined periodically and NO_3-N was determined at the end of the experiment.

St. Petersburg Compost

Greenhouse – A commercial product packaged from composted municipal refuse under the tradename of Cura was obtained from the compost plant at St. Petersburg, Florida, for this experiment. The compost (Table 6.1) was mixed with Arredondo sand which had pH 5.8 and contained 12 ppm P, 28 ppm K, 388 ppm Ca, and 73 ppm Mg. The soil was limed at the rate of 2 metric tons per hectare and weighed into portions of 4,500 grams each. The soil was mixed thoroughly with compost at rates which progressed geometrically from 1 metric ton per hectare through 64 metric tons per hectare, placed in pots, and arranged in completely randomized blocks of four replications in the greenhouse.

Two control treatments which contained no added fertilizer or a 10-10-10 fertilizer applied at 1 metric ton per hectare were used for comparison. The total number of treatments was nine. The soil was wet thoroughly with distilled water and allowed to drain for two days. The pots were weighed and an average weight was obtained as a guide for watering throughout the experiment.

Cranberry beans, *Phaseolus vulgaris* Savi, were planted, thinned to three plants per pot at 14 days from planting, and harvested at 45 days from planting. Oats, *Avena sativa* L., were planted in the pots, after the beans were removed, and harvested at 70 days from planting. Additional fertilizer and compost were applied at the original rates to the appropriate pots, and three 14-day old tomato, *Licopersicon esculentum* Mill., seedlings were transplanted from vermiculite into each pot. Two of these plants were harvested at 14 days from transplanting and the remaining plant was harvested at 56 days from transplanting. The plant tissue was dried at 70°C, weighed, ground to pass a 20-mesh screen, and analyzed chemically. The soil was sampled after the tomato plant harvest and analyzed chemically.

Fairfield Compost

Greenhouse – A pelletized compost manufactured by the Fairfield Engineering Company, Marion, Ohio, was obtained for this experiment. Arredondo sand with pH 6.6 and containing 11 ppm P, 12 ppm K, 565 ppm Ca, and 6 ppm Mg was weighed in 6,000 gram portions and mixed with compost (Table 6.1) at rates of 8 to 64 metric tons/ha. Again, control treatments with no added fertilizer or 2 tons/ha of 10-10-10 fertilizer were used as comparisons.

Sorghum, *Sorghum vulgare* Pers., seeds were planted, thinned to four plants per pot at 14 days from planting, and harvested at 70 days from planting. Compost and fertilizer were added again at the original rates to the appropriate pots, sorghum seeds planted, thinned to 6 plants per pot, and harvested at 45 days from planting. Plant tissue was dried at 70°C, weighed, ground to pass a 20-mesh screen, and analyzed chemically.

Gainesville Compost

Field – Compost prepared by the Gainesville Municipal Waste Conversion Authority, Gainesville, Florida, was used in this experiment. The compost (Table 6.1) was applied to plots (3 x 7 meters) located on phosphate sand tailings at International Minerals and Chemical Corporation, Bartow, Florida. The experimental design was a 5 x 5 latain square with treatments as follows:

(1) 1 ton/ha of 10-10-10 fertilizer,
(2) 35 tons/ha of compost plus 10-10-10,
(3) 70 tons/ha of compost plus 10-10-10,
(4) 35 tons/ha of compost, and
(5) 70 tons/ha of compost.

Compost and fertilizer were applied in the spring of two consecutive years and sorghum was planted. Sorghum was harvested, fertilizer only reapplied at 1 ton per hectare, and oats were planted in the fall of the two years. Plant yields, tissue analyses, and soil analyses were recorded as in other experiments.

RESULTS AND DISCUSSION

Largo Compost

Greenhouse – The specific conductance of the water extracts increased greatly as the amount of compost extracted increased 1.97, 3.50, 5.85, 11.00, 18.00, and 26.60 millimhos per cm for the 10, 20, 40, 80, 160, and 320 gram samples, respectively. Above 8 millimhos/cm is considered a strongly saline solution and only tolerant plant species will grow satisfactorily in a soil that produces an extract of this concentration or higher.

Of the plant species chosen for this study, turnip has good salt tolerance, oat and pearl millet plants have moderate salt tolerance. The germination of oat and pearl millet seeds (Table 6.2) was not significantly depressed except by the extract from the highest compost rate (320 grams) and, in fact, there was improved germination where extracts from 40 to 80 grams of compost were used.

TABLE 6.2: EFFECTS OF COMPOST/WATER EXTRACTS ON SEED GERMINATION AND SEEDLING LENGTH

Compost	Oats		Turnip		Radish		Millet	
g	%	mm	%	mm	%	mm	%	mm
0	92	123	96	48	96	72	82	74
10	94	151	100	82	100	100	82	90
20	92	149	96	91	100	98	80	80
40	98	151	93	87	96	90	91	91
80	99	131	84	77	56	82	88	99
160	96	74	40	14	16	17	78	70
320	41	29	0	0	0	0	45	37
L.S.D. 5%	5	20	15	7	5	18	5	12
1%	7	25	17	9	7	20	7	16

However, turnip seeds under the conditions of this experiment were not tolerant of the two highest rates of compost and radish seed germination was depressed greatly at the 80 gram compost rate.

Seedling growth of all species increased in the extracts from 10 to 80 grams of compost as compared to distilled water. In the 160 gram compost extract no radicles developed on turnip seedlings. The extracts were not analyzed for plant nutrients and no attempt was made to determine if a phytotoxic compound was present so no explanation is offered for these beneficial or detrimental effects.

Oat foliage yields and nutrient uptake are presented in Table 6.3. The highest yield was from plants grown in the 512-ton compost treatment. Nitrogen uptake by the oat plants was much higher from the 512-ton compost treatment than from the other treatments. The lack of significant increase in N uptake by oat plants growing in the 2, 8, and 32 tons/ha treatments indicated that N in the compost was probably diverted to supply the needs of the microbial population. Phosphorus uptake was significantly higher in plants from the two highest rates of compost. As with N uptake, K uptake was greatest in the highest compost rate.

TABLE 6.3: YIELDS AND N, P, AND K UPTAKE BY OAT FOLIAGE

Material applied	T/ha	Oven-dry	N	P	K
		g/pot	----------mg/pot--------		
None	0	1.6	57	3	92
Compost	2	2.2	66	3	88
Compost	8	2.6	61	3	112
Compost	32	1.9	63	4	104
Compost	128	2.4	87	6	128
Compost	512	3.4	142	9	173
L.S.D.	5%	0.5	20	1	35
	1%	0.7	30	2	42

Turnip foliage growth (Table 6.4) was increased greatly in the three highest compost treatments. Nitrogen deficiency symptoms, light green color, changing to yellow with age, appeared in all of the other three treatments. On the other hand, turnip plants growing in the 512-ton compost treatment exhibited striking phytotoxic effects which were not identified. The symptoms were not the usual stunting and blue-green color attributable to salt injury, but there was a spiraling in the older leaves which progressed until the leaf cohered into a thin, stem-like appendage.

TABLE 6.4: YIELDS AND N, P, K UPTAKE BY TURNIP FOLIAGE

Material applied	T/ha	Oven-Dry	N	P	K
		g/pot	----------mg/pot---------		
None	0	1.7	34	5	31
Compost	2	2.9	54	6	50
Compost	8	2.7	54	7	54
Compost	32	4.3	86	12	103
Compost	128	6.0	113	21	168
Compost	512	5.2	282	18	295
L.S.D.	5%	1.5	25	3	25
	1%	2.0	30	5	30

There was no discoloration in the affected leaves or in the rest of the plant. Nitrogen, P, and K uptake was significantly greater in the turnip plants grown in the three highest rates of compost. Increased yield and N uptake by turnip plants grown in the 32-ton compost treatment indicated that the soil microbial population was releasing N for higher plants utilization. At that time the compost had been in the soil for 6 months.

Radish foliage yields (Table 6.5) were significantly higher in the 128-and 512-ton compost treatments. Yields in the 2-and 32-ton compost treatments were significantly lower than the control and the 8-ton treatment yielded the same as the control. Here again, N deficiency symptoms appeared in the control and the three lowest compost treatments.

There were no indications of phytotoxic effects maifested by any radish plant. The uptake of N, P, and K was significantly greater in plants grown in the 128- and 512-ton compost treatments as compared to plants grown in the other treatments. The low N uptake by plants from the three lowest compost treatments was further evidence of microbial competition for available N.

TABLE 6.5: YIELDS AND N, P, K UPTAKE BY RADISH FOLIAGE

Material applied	T/ha	Oven-dry	N	P	K
		g/pot	--------mg/pot--------		
None	0	4.8	123	8	89
Compost	2	3.9	112	6	56
Compost	8	4.2	103	8	90
Compost	32	3.1	87	8	133
Compost	128	7.5	192	18	355
Compost	512	7.0	237	22	367
L.S.D.	5%	1.4	30	5	125
	1%	1.8	35	6	150

Pearl millet yields (Table 6.6) in the 512-ton compost treatment were over two times the 128-ton compost treatment yields. There were N deficiency symptoms manifested by plants growing in the control and 2-ton compost treatments; however, none of the symptoms was severe. There were no recognizable phytotoxic symptoms manifested by any of the pearl millet plants growing in the 512-ton compost treatment.

Potassium uptake was of particular interest, as the plants in that treatment removed over 50 times the amount of K removed by the control plants. Potassium deficiency should have developed in the plants growing in the control pots; however, no symptoms were manifested.

As shown in Tables 6.7 and 6.8, the two highest rates (128 and 512 tons/ha) of compost increased significantly soil P, K, Ca, Mg, TSS, pH, CEC, and water-holding capacity. In addition, the soil at the end of this experiment appeared to have better structure where the two highest rates of compost were added.

TABLE 6.6: YIELDS AND N, P, K UPTAKE BY PEARL MILLET FOLIAGE

Material applied	T/ha	Oven-dry	N	P	K
		g/pot	----------mg/pot--------		
None	0	7.8	107	12	23
Compost	2	7.6	113	10	45
Compost	8	9.2	125	16	34
Compost	32	11.8	146	16	86
Compost	128	18.2	203	37	597
Compost	512	39.1	305	57	1,230
L.S.D.	5%	6.0	65	8	67
	1%	8.5	77	10	79

TABLE 6.7: EFFECTS OF TREATMENT ON SOIL

Material applied	T/ha	P	K	Ca	Mg	TSS	pH
		-------------------ppm-------------------					
None	0	1	7	738	16	287	6.2
Compost	2	3	9	669	16	229	6.0
Compost	8	3	8	684	16	229	6.0
Compost	32	3	9	882	16	258	6.1
Compost	128	6	13	1,165	42	607	6.4
Compost	512	10	17	1,682	79	1,512	6.5
L.S.D.	5%	2	4	158	11	131	.1
	1%	3	5	215	15	178	.2

TABLE 6.8: EFFECTS OF TREATMENT ON CATION EXCHANGE CAPACITY AND WATER-HOLDING CAPACITY

Material applied	T/ha	CEC	Soil Moisture (%)		
		meq/100 g	1/10 atm	1/3 atm	15 atm
None	0	3.67	6.42	4.28	3.84
Compost	2	3.70	6.45	4.18	4.00
Compost	8	3.78	6.65	4.39	3.96
Compost	32	3.90	7.34	4.78	4.36
Compost	128	4.81	8.15	5.91	5.52
Compost	512	7.14	10.79	8.62	8.10
L.S.D.	5%	.48	.47	.37	.35
	1%	.64	.63	.50	.46

Field — The oat seeds germinated normally in all pots; however, seeds in soil containing the highest rate of compost (512 metric tons/ha) were from 3 to 5 days later than the other seeds in germinating. For the first few weeks, oat seedlings in all pots grew satisfactorily with no indications of nutrient deficiencies. When the seedlings were 6 to 7 weeks old, N deficiency symptoms, yellowish green to yellow leaves and purplish green stems, began to appear in the control pots and the symptoms increased in severity as growth continued.

In addition, N deficiency sysmptoms were observed after the first cutting (at 12 weeks of age) in the pots containing 2 and 8 tons/ha of compost; however, plants in those pots never developed the severe N deficiency symptoms noted in the control pots. No visual evidence of other nutrient deficiencies were discernible, but other deficiency symptoms would likely have been masked by the N deficiency symptoms.

Total yields of oat foliage from two cuttings (Table 6.9) were increased significantly over the control by all treatments, and yields increased progressively within the compost treated soil as the amount of compost applied increased. The highest yield (52 grams) was obtained from the highest rate of compost. Plant growth was, no doubt, a function of the added N available to the plants (168 mg in the lowest rate of compost up to 43 grams in the highest rate, and 12.23, 2.96, 5.91 and 1.40 grams in the sludge, cow manure, chicken manure, and 10-10-10 treatments, respectively).

Nitrogen uptake by the oat plants (Table 6.9) was significantly increased by all treatments except where the three lowest rates of compost were applied. Frequently, plants suffering under a deficiency of a particular nutrient will contain a relatively high concentration of that nutrient. This was true of the oat plants from the control pots which contained 2.58% N (average of two cuttings) as compared to 2.29, 2.07, 1.83, 2.04, 2.54, 3.55, 1.86, 3.27 and 2.24% N in plants from the other pots, as listed from top to bottom in Table 6.9.

The decrease in N content of plants grown in the first three compost treatments may have resulted from microbial competition as soil microbes satisfy their needs before higher plant species. With regard to other nutrients, only the plants grown in the two lowest rates of compost failed to show significantly higher uptake than the control plants with the noticeable exception of K and Mn uptake from the sludge treatment, which was quite low in both K and Mn.

Yields of fresh radish roots and oven-dry tops (Table 6.10) were greatly increased where the highest rate of compost was applied. Sewage sludge and chicken manure were the only other materials that produced significantly higher yields than the control. In fact, radish plants from all of the other treatments, except cow manure, yielded lower than control plants. A large part of this negative response was doubtless caused by the removal of N by the oat plants. In addition, some added N was certainly lost through leaching. Since neither explanation applied fully to the reduced yields in pots that had received 128 tons/ha of compost, increased microbial activity seemed to deprive the radish plants of necessary N.

TABLE 6.9: INFLUENCE OF TREATMENT ON YIELD AND NUTRIENT UPTAKE BY OATS

Material	Amount applied	Yield	N	P	K	Ca	Mg	Mn	B	Cu	Zn
	Metric tons/ha	g/pot	---	---	---	---	mg/pot	---	---	---	---
Garbage Compost	2	26	604	36	435	134	58	1.4	0.3	0.4	1.7
" "	8	28	594	39	560	135	57	1.8	0.3	0.5	1.8
" "	32	36	678	53	1015	166	80	5.1	0.5	0.6	3.1
" "	128	49	1011	116	1865	231	126	14.9	1.1	1.0	5.7
" "	512	52	1170	101	1880	245	110	16.0	2.6	1.1	4.4
Sewage sludge	32	42	1405	157	413	375	112	2.9	0.5	0.9	4.4
Cow manure	32	45	896	138	1955	151	89	3.4	0.4	0.6	4.0
Chicken manure	32	45	1443	150	1736	234	100	5.0	0.6	0.6	4.0
10-10-10+ Mg, Mn, Cu, Zn, B	2	46	1051	76	1000	218	116	8.6	0.7	0.9	5.4
Control	-	19	494	22	215	111	39	0.8	0.2	0.3	1.0
L.S.D.	5%	6	229	23	432	50	26	3.5	0.4	0.2	1.4
	1%	8	311	32	590	68	35	4.8	0.6	0.3	1.9

TABLE 6.10: INFLUENCE OF TREATMENT ON RADISH YIELDS AND NUTRIENT UPTAKE BY RADISH TOPS

Material	Amount applied	Yield: Fresh roots	Yield: Oven-dry tops	N	P	K	Mn	Zn	Cu
	Metric tons/ha	------g/pot-------		-------------------mg/pot-------------------					
Garbage Compost	2	27	2.0	56	56	24	.05	0.2	.02
" "	8	36	2.3	59	62	25	.07	0.3	.02
" "	32	34	1.9	50	55	25	.07	0.2	.02
" "	128	39	2.1	62	84	47	.11	0.3	.02
" "	512	187	9.4	375	263	467	.54	3.0	.16
Sewage sludge	32	113	8.9	260	695	65	.43	1.7	.12
Cow manure	32	67	3.2	99	126	61	.17	0.3	.04
Chicken manure	32	151	6.6	204	304	126	.33	0.7	.07
10-10-10+Mg, Mn, Cu, Zn, B	2	43	2.7	66	82	30	.17	0.4	.03
Control	-	50	3.6	100	105	35	.08	0.5	.04
L.S.D.	5%	37	1.0	37	72	79	.10	0.3	.03
	1%	50	1.3	51	90	108	.13	0.4	.04

The uptake of plant nutrients by radish tops (Table 6.10) followed the same trend as yields where significant differences were concerned. However, there was one outstanding exception in that K uptake was greatly curtailed where sewage sludge was applied. The low K content (0.9%) in plants from the sewage sludge treatment, as compared to 0.16% K in control plants, indicated that there was a soil K deficiency.

Typical N deficiency symptoms, small, yellow leaves and small roots, were prevalent in the first four rates of compost and in the cow manure, 10-10-10, and control treatments. In addition, N uptake by radish plants in those treatments was further evidence of a deficient N supply.

The effects of treatment on some soil measurements (Table 6.11) were interesting in several respects. Soil pH averaged 6.9 but as pH was not significantly affected by treatment those data are not included. Water-holding capacity, as measured by moisture equivalents, was increased significantly only by the 128- and 512-ton rates of compost, whereas cation exchange capacity was increased significantly only by the 512-ton rate of compost.

The two highest total soluble salts and Ca contents were measured in soil which had received 128 and 512 tons/ha of compost. Phosphorus content was increased significantly over the control by all treatments except the two lowest compost rates. The relatively low K content of sewage sludge was again emphasized by the small increase in soil K.

It was quite evident through visual observation, plant yield measurements, and plant and soil analyses that the 512-ton rate of garbage compost was superior to all other treatments. However, such a high rate, especially in an initial application, is uneconomical except for a few speciality crops or on areas as golf courses, lawns, home gardens, etc.

Microbiological — Figure 6.1 shows the relative numbers of bacteria and actinomyces per gram of a mixture of finely ground garbage compost and Arredondo fine sand. In the soil microbial population studies, a rapid increase in relative bacterial numbers was obtained during the first 2 days. A rapid drop in bacterial numbers occurred by the 4th day (Figure 6.1).

This drop was followed by a general leveling of the population in the 1-and 5-gram treatments. The same initial trend occurred in the 10-gram treatment. However, the numbers increased again by the 6th day, then gradually decreased.

Figure 6.2 shows the total amount of CO_2-C evolved from mixtures of Arredondo fine sand and organic materials over a period of 65 days. An increase in CO_2 evolution was obtained as the levels of each organic material were increased. This increase in CO_2 evolution was not the same for each of the four materials. The rate of increase was faster for sewage sludge and chicken manure and slower for cow manure. There was slightly less CO_2 evolved from chicken manure than from sewage sludge. This was true at all levels.

TABLE 6.11: INFLUENCE OF TREATMENT ON PHYSICAL AND CHEMICAL PROPERTIES OF LEON FINE SAND

Material	Amount applied	ME	CEC	TSS	Ca	P	K	Mg
	Metric tons/ha	%	meq/100 g	ppm				
Garbage Compost	2	5.19	4.05	272	825	2.0	10	26
" "	8	5.33	4.01	297	860	2.0	15	24
" "	32	5.65	4.27	569	915	3.5	19	32
" "	128	6.98	4.90	1035	1188	7.5	95	54
" "	512	11.54	8.28	1663	1981	11.0	223	104
Sewage sludge	32	5.57	4.51	256	690	11.0	12	38
Cow manure	32	5.73	4.27	674	986	12.3	133	60
Chicken manure	32	5.33	4.55	296	895	16.5	23	60
10-10-10 + Mg, Mn, Cu, Zn, B	2	5.60	4.84	641	766	4.0	12	39
Control	-	5.83	4.67	171	785	1.8	10	20
L.S.D.	5%	.41	.66	330	140	0.4	60	14
	1%	.55	.90	448	190	0.5	81	19

FIGURE 6.1: BACTERIA AND ACTINOMYCES IN COMPOST-SAND MIXTURE

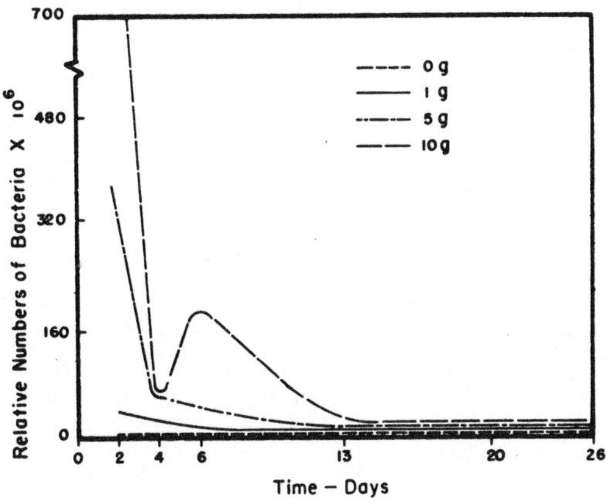

FIGURE 6.2: CO_2-C EVOLUTION FROM SAND-ORGANIC MATTER

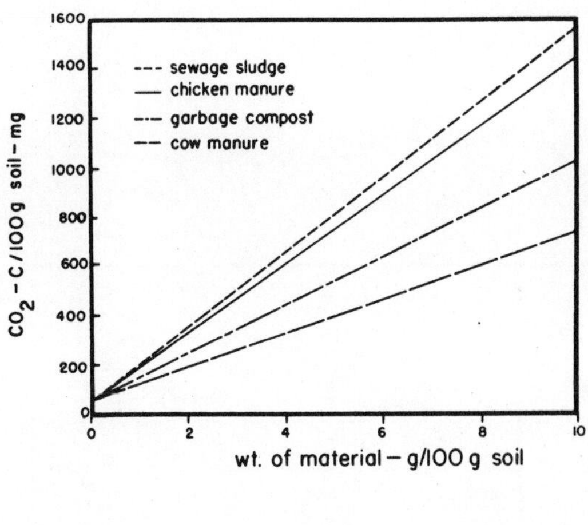

Carbon dioxide evolved from chicken manure and sewage sludge at the 10-gram level was 99 and 115% greater than cow manure and 44 and 55% greater than garbage compost. At this same level, CO_2 evolution from garbage compost was 28% more than cow manure. Municipal refuse contains a large percentage of cellulose, sugars, and starches and sugars and starches should be readily decomposed by microorganisms.

In animals, the quality of feed, age, and condition of the animal influence the amount of cellulose or crude fiber left in the feces. Because partial decomposition has taken place in the animal, a larger percentage of more resistant compounds should exist. Therefore, one would expect cow manure to be more resistant to soil microbial attack than garbage compost. This should be reflected in a reduction of CO_2 evolved. Results obtained indicated significantly less CO_2 was evolved from treatments containing cow manure than garbage compost. On the other hand, chicken manure and sewage sludge contain a higher percentage of N and easily decomposable compounds. Thus one would expect these materials to decompose readily after being added to soil. When total CO_2 evolution from the treatments was compared, a highly significant interaction was observed between materials and level applied.

Separate regression equations were obtained for each material and the slopes of regressions lines were significantly different. The relationship between treatment levels and total CO_2 evolved was linear in all cases. There was no significant difference detected in total CO_2 for all levels used in this study for chicken manure and sewage sludge. At all levels, CO_2 production for chicken manure and sewage sludge was significantly greater than that obtained with cow manure or garbage compost, there was no significant difference at the lower levels. However, garbage compost became significantly different from cow manure between the 6- and 8-gram level.

Figure 6.3 shows the NO_3-N produced from mixtures of Arrendondo fine sand and organic materials after 65 days. At the 2-gram level, the amount of NO_3-N produced was from highest to lowest chicken manure, sewage sludge, cow manure, and garbage compost.

FIGURE 6.3: NO_3-N PRODUCTION

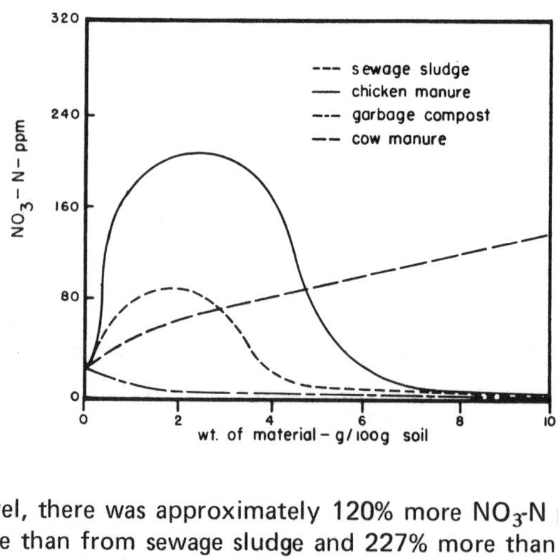

At this low level, there was approximately 120% more NO_3-N produced from chicken manure than from sewage sludge and 227% more than from cow manure. However, a rapid decrease occurred in NO_3-N production when levels of chicken manure and sewage sludge were increased. This decrease was probably the result of an inhibitory effect on nitrification by the high levels of N added. At the 2-gram level, 528 and 1,092 ppm N, respectively, were added when chicken manure and sewage sludge were used. Nitrate nitrogen in the cow manure treatments increased linearly as levels of material were increased. Very little nitrification occurred in the garbage compost treatments. This was probably due to rapid immobilization of N by microflora during decomposition of the compost. A highly significant interaction was obtained between materials and level applied.

In the study involving 11 levels of garbage compost, there was a progressive increase in amounts of CO_2 evolved as the amount of compost was increased. Figure 6.4 shows the total amount of CO_2-C evolved from ground garbage compost mixed with Arredondo fine sand over a period of 101 days.

FIGURE 6.4: CO_2-C EVOLUTION FROM COMPOST-SAND MIXTURES

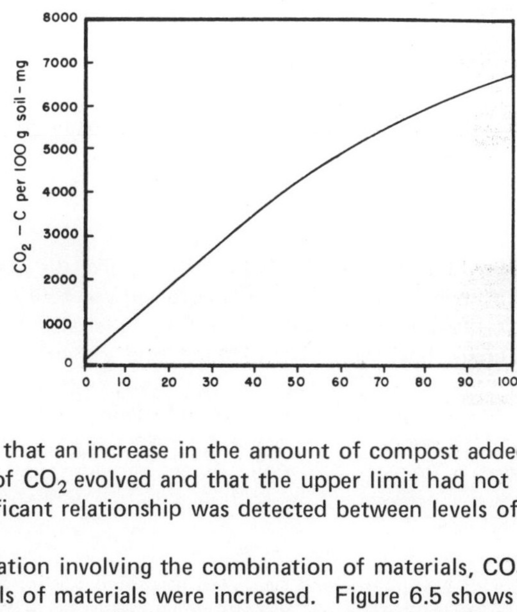

It was evident that an increase in the amount of compost added increased the total amount of CO_2 evolved and that the upper limit had not been reached. A highly significant relationship was detected between levels of compost used.

In the investigation involving the combination of materials, CO_2 evolution increased as levels of materials were increased. Figure 6.5 shows the total amount of CO_2-C evolved over a 64-day period from mixtures of finely ground garbage compost with cow manure, chicken manure and sewage sludge mixed with Arredondo fine sand. This was true for all treatments. As in the single material study, the rate of increase was faster for sewage sludge and chicken manure and slower for cow manure.

At the highest level of application, CO_2 evolution from chicken manure and sewage sludge was 30 and 55%, respectively, more than cow manure. An interesting observation was noted when the results of this study were compared with results obtained from the single material study (Figure 6.2). When CO_2 values from the garbage compost were added to those of cow manure, chicken manure, or sewage sludge and compared with those obtained in this study, very little difference was obtained. This indicated that combining materials had little effect on rate of decomposition.

FIGURE 6.5: CO_2-C EVOLUTION FROM COMPOST-ORGANIC MATTER-SAND MIXTURES.

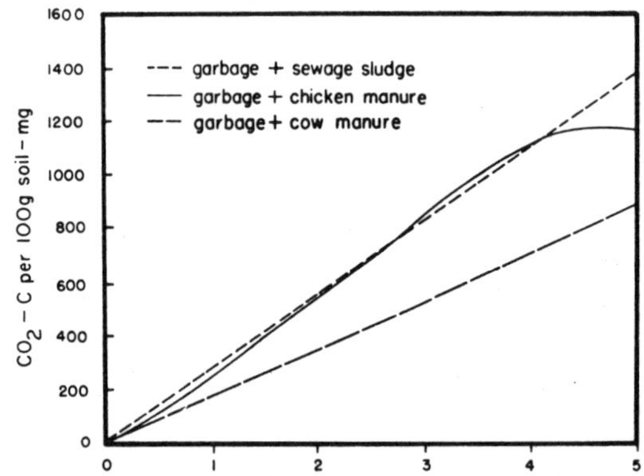

Analyses of variance indicated that for total CO_2 evolved, the effects of material and rate are independent. The total CO_2 evolved was found to be significantly less for cow manure than for the other two materials. No significant difference was detected between chicken manure and sewage sludge.

Figure 6.6 shows the NO_3-N produced at the end of 65 days from combinations of ground garbage compost with cow manure, chicken manure and sewage sludge mixed with Arredondo fine sand. For NO_3-N, the relationship of chicken manure to sewage sludge was somewhat similar to that observed with single materials (Figure 6.3).

However, production was greater and occurred at a higher level than with single materials. This was the reverse from the results obtained in the second study (Figure 6.3) which indicated that the available N in the cow manure may have been immobilized by the microflora in the decomposition of the combined materials.

These results also illustrated the modifying influence of compost in reducing the excessive amounts of nitrogen available for nitrification. For NO_3-N the interaction between levels of materials was found to be significant. In general, the relative number of fungi in garbage compost increased with time of incubation and bacteria decreased with time of incubation.

FIGURE 6.6: NO_3-N PRODUCTION FROM GARBAGE-ORGANIC MATTER-SAND MIXTURES

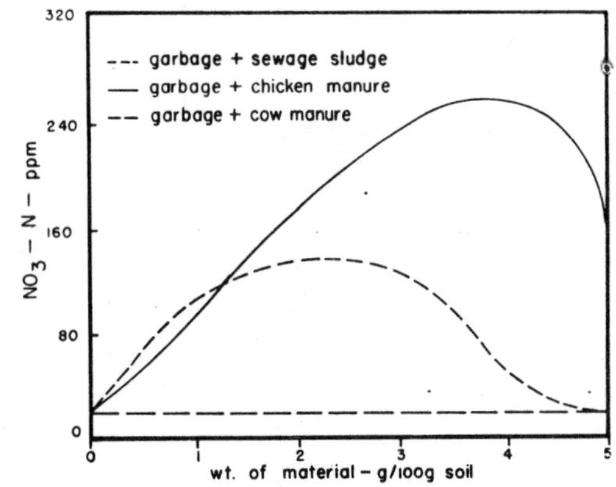

Little nitrification occurred in the garbage compost treatments; nitrification rate decreased rapidly as levels of chicken manure and sewage sludge were increased above the 2-gram level; chicken manure and sewage sludge decomposed at a faster rate than garbage compost or cow manure; and garbage compost decomposed at a faster rate than cow manure.

St. Petersburg Compost

Greenhouse – Cranberry beans grew normally in all of the treatments with no evidence of nutrient deficiences or phytotoxic effects in any plants. There were no significant differences among treatments with respect to seedling weights (Table 6.12). However, compost applied at the 16-ton rate and higher resulted in significantly more bean pods and fresh bean weights than either control treatment.

TABLE 6.12: YIELDS OF BEAN PLANT MATERIAL

Material	Rate	Seedlings	Pods/pot	Fresh Beans	Oven-dry Plants
	tons/ha	g/pot	No./pot	g/pot	g/pot
Cura	1	4.0	6	9.8	5.9
"	2	4.2	8	15.8	6.3
"	4	4.4	7	16.2	6.7
"	8	4.4	6	14.4	7.0
"	16	4.9	9	21.2	8.2
"	32	5.0	10	18.9	7.7
"	64	6.0	11	20.4	8.6
10-10-10	1	4.9	5	12.7	6.2
None	-	5.0	6	13.5	6.2
L.S.D.	5%	N.S.	3	5.0	2.1
	1%	N.S.	4	6.7	N.S.

Uptake of N and K by bean seedlings was not significantly affected by treatment (Table 6.13); however, K uptake was significantly higher in seedlings removed from the 64-ton rate as compared to the controls.

TABLE 6.13: UPTAKE OF N, P, AND K BY IMMATURE BEAN PLANTS

Materials	Rate	N	P	K
	Tons/ha		-mg/pot-	
Cura	1	197	28	104
"	2	199	28	97
"	4	220	31	108
"	8	223	32	121
"	16	240	34	147
"	32	238	30	149
"	64	296	34	262
10-10-10	1	253	34	148
None	-	207	31	101
L.S.D.	5%	N.S.	N.S.	58
	1%	N.S.	N.S.	79

In contrast, mature beans removed significantly greater amounts of N-P-K from soil treated with 16 tons per hectare and higher of compost (Table 6.14).

TABLE 6.14: UPTAKE OF N, P, and K BY MATURE BEAN PLANTS

Material	Rate	N	P	K
	Tons/ha		-mg/pot-	
Cura	1	104	14	76
"	2	105	17	78
"	4	117	21	83
"	8	111	22	105
"	16	155	25	150
"	32	126	22	146
"	64	147	25	225
10-10-10	1	112	15	95
None	-	90	15	58
L.S.D.	5%	38	8	34
	1%	N.S.	N.S.	46

The oat plants grew normally with no indications of nutrient deficiencies, but yields and N-P-K uptake (Table 6.15) showed evidence that these elements were in short supply in the control (10-10-10 and no fertilizer) pots. Oat plants growing in the presence of 16 tons/ha and higher of compost significantly outyielded the controls and N-P-K uptake was much greater from the 32-and 64-ton compost applications than from the controls. It was quite evident that the residual compost supplied adequate amounts of N-P-K for the oat plants and at the two highest rates (32 and 64 tons/ha) there was luxury consumption of K.

TABLE 6.15: YIELDS AND UPTAKE OF N, P, AND K BY OAT PLANTS

Materials	Rate	Oven-dry	N	P	K
	Tons/ha	g/pot	----------	mg/pot	----------
Cura	1	6.7	70	26	92
"	2	6.3	68	26	84
"	4	6.0	61	27	75
"	8	6.4	66	30	101
"	16	7.8	70	33	101
"	32	8.2	92	40	179
"	64	8.1	93	44	225
10-10-10	1	5.7	51	26	92
None	-	5.5	62	25	70
L.S.D.	5%	1.1	13	4	25
	1%	1.4	18	6	34

After the oat plants were harvested, additional compost or 10-10-10 fertilizer was added in the original amount. After 7 days, tomato seedlings transplanted from vermiculite were well established and growing in all pots except where 32 and 64 tons/ha of compost were applied. The seedlings in the 32-ton treatment survived, but did not start growing until 21 days from transplanting.

In the 64-ton treated soil three seedlings died and were replaced, and growth did not resume in those pots until the seedlings had been in the soil for over 30 days. This compost contained some chemical that was highly toxic to tomato; however, the responsible toxicant was not identified.

Yields of tomato plant material (Table 6.16) from the 10-10-10 control pots were over twice the yields from any other pot. The yields from the 64-ton rate of compost were reduced drastically as indicated above. The only compost treatment that outyielded significantly the no-fertilizer control was the 8-ton rate.

TABLE 6.16: YIELDS AND UPTAKE OF N, P, AND K BY TOMATO PLANTS

Material	Rate	Oven-dry	N	P	K
	Tons/ha	g/pot	----------	mg/pot	----------
Cura	1	6.1	96	32	62
"	2	6.4	90	21	64
"	4	6.2	90	22	70
"	8	7.6	101	25	113
"	16	4.9	67	22	118
"	32	4.7	55	24	151
"	64	1.9	29	10	69
10-10-10	1	15.2	197	34	227
None	-	5.8	97	20	50
L.S.D.		1.7	18	7	23
		2.4	25	9	31

Uptake of N-P-K was much greater from the mineral fertilizer than from the compost. There was no doubt that the tomato plants in the no-fertilizer control suffered from K deficiency as those plants contained only 0.85% K which waş definitely below the deficiency level.

Compost applications to Arredondo sand increased soil pH, as compared to the no-fertilizer control, whereas 10-10-10 fertilizer decreased the pH (Table 6.17). Soluble salts were increased by the two highest compost applications (32 and 64 tons/ha); however, it is doubtful that these levels were high enough to be harmful to plant life.

TABLE 6.17: INFLUENCE OF TREATMENT ON CHEMICAL PROPERTIES OF ARREDONDO SAND

Material	Rate	pH	TSS	P	K	Ca	mg
	Tons/ha		ppm				
Cura	1	6.60	415	8	97	751	77
"	2	6.60	394	8	131	918	86
"	4	6.60	545	12	235	1315	105
"	8	6.65	435	15	322	1645	115
"	16	6.70	495	19	365	1865	128
"	32	6.60	746	21	442	2160	144
"	64	6.80	1288	25	590	2798	191
10-10-10	1	6.05	543	7	41	516	68
None	-	6.50	417	6	12	354	59
L.S.D.	-	0.10	189	2	49	235	25
	-	0.13	257	3	66	320	34

The TSS levels were surely much higher in all treatments soon after compost or fertilizer applications, but several liters of water added for plant use had moved the salts out of the soil. Soil levels of P, K, Ca, and Mg were greatly increased by two compost applications of 4 to 64 tons/ha. Potassium levels in the control pots were much lower than in any of the compost treated pots.

Fairfield Compost

Greenhouse – The first sorghum seeds planted germinated satisfactorily in pots and the plants grew normally without manifesting any deficient or toxic symptoms. Sorghum plants that were growing in soil containing the two highest rates of compost (32 and 64 tons/ha) and 10-10-10 fertilizer produced seedheads 7 days before the two lowest rates of compost (8 and 16 tons/ha) and 14 days before the no-fertilizer control. Average plant height at maturity was greatest in sorghum from the 32-ton rate of compost (704 mm) and least from the no-fertilizer control (379 mm).

Sorghum yields were significantly increased by all compost applications as compared to no fertilizer (Table 6.18). However, only the 64-ton compost rate outyielded the 10-10-10 fertilizer control. Uptake of N-P-K was also significantly greater in plants grown where compost was applied as compared to no fertilizer. Uptake of N and K from the compost treated soil was almost linear.

TABLE 6.18: YIELDS AND N, P, AND K UPTAKE BY THE FIRST SORGHUM PLANTS

Materials	Rate	Oven-dry	N	P	K
	Tons/ha	g/pot	----------mg/pot--------		
None	-	11.2	102	33	55
Compost	8	16.7	194	58	85
"	16	20.6	269	71	117
"	32	24.8	383	79	171
"	64	26.0	579	77	254
10-10-10	2	22.3	412	67	368
L.S.D.	5%	3.0	78	9	31
	1%	4.2	108	13	43

The second sorghum crop was normal in all treatments except where no fertilizer or compost was applied. As shown in Table 6.19, yields of sorghum plants grown in the unfertilized soil were much lower than yields from the other treatments and greatly reduced as compared to the first crop (1.4 grams versus 11.2 grams).

TABLE 6.19: YIELDS AND N, P, AND K UPTAKE BY THE SECOND SORGHUM PLANTS

Material	Rate	Oven-dry	N	P	K
	Tons/ha	g/pot	---------mg/pot------		
None	-	1.4	38	7	5
Compost	8	12.1	124	40	48
"	16	18.9	183	47	86
"	32	31.5	312	68	178
"	64	39.6	673	82	352
10-10-10	2	34.4	504	62	585
L.S.D.	5%	3.5	82	9	21
	1%	4.9	114	13	29

This indicated that the first crop removed a large portion of the available plant nutrients from the soil leaving very little for the second crop. On the other hand, residual plant nutrients plus the additional compost or fertilizer greatly benefited plants growing in the other treatments as shown by plant yields and N-P-K uptake.

Gainesville Compost

Field – The phosphate mining process involves a procedure in which the phosphate is removed from the matrix through flotation. After the phosphate is removed, there remain two waste products, phosphate slimes and sand tailings, which are of major concern to the industry as they present tremendous problems of disposal and land reclamation. The sand tailings are pumped onto mined out areas to depths of 20 feet and higher, allowed to drain, and planted to suitable vegetation. This material is a coarse sand, quite low in available plant nutrients (pH 7.5, and containing 36 ppm P, 5 ppm K, 223 ppm Ca, and 36 ppm Mg extracted in NH_4OAc at pH 4.8), and with little cation exchange capacity or water-holding capacity.

The first crop, sorghum, planted in this experiment was slow in becoming established and showed evidence of nutrient deficiencies almost continuously from seedling stage to maturity. Torrential rains, during the 4th and 5th weeks from planting induced severe N deficiency symptoms in all plots. Therefore, NH_4NO_3 was applied as a sidedressing to the sorghum growing in plots which received 10-10-10. Seedhead yields in this first crop (Table 6.20) were significantly increased where compost plus 10-10-10 was applied as compared to 10-10-10 alone.

TABLE 6.20: YIELDS OF SORGHUM SEEDHEADS AND OAT FOILAGE

Compost	10-10-10	Sorghum (g/plot)		Oats (g/plot)	
----Tons/ha-----		1968-'69	1969-'70	1968-'69	1969-'70
0	1	290	626	2086	1022
35	1	492	886	2723	1520
70	1	540	1248	2674	1838
30	0	8	151	0	86
70	0	28	432	0	130
L.S.D.	5%	124	118	387	304
	1%	174	174	542	424

However, yields from plots which received only compost were negligible. Yields from the first crop of oats were also increased greatly where compost plus 10-10-10 was applied. Oat seeds germinated satisfactorily in the plots containing residual compost only, but 100% of the seedlings died shortly after germination. The cause of death was not determined; however, it was doubtless brought about by severe N deficiency. The second crop of sorghum appeared much healthier than the first crop and seedhead yields were much greater. Also, the second oat crop was much better than the first crop.

The large increase in sorghum seedhead yields from all plots in the second year as compared to the first year was outstanding. This was particularly noteworthy in the compost only plots. Oat seeds were broadcast on the plots in the first year and were eaten by redwing blackbirds which necessitated replanting. Oat seeds were planted in furrows/plot, 72 cm apart, in the second year.

Therefore, yields cannot be compared between years, but the second year yields would certainly have been 3 or 4 times higher under broadcast planting. Again with oats, the increase in yields from compost only plots between years was noteworthy. Bird pilferage during both years prevented oat seed harvests and forced the planting of bird resistant sorghum seed.

The addition of compost plus 10-10-10 to sand tailings resulted in increased K contents of sorghum the first year and increased N-K contents the second year as compared to 10-10-10 alone (Table 6.21).

TABLE 6.21: EFFECTS OF COMPOST ON N-P-K CONTENTS OF SORGHUM PLANTS

Compost	10-10-10	1968-'69 (%)			1969-'70 (%)		
----Tons/ha------		N	P	K	N	P	K
0	1	1.51	0.55	0.39	0.82	0.40	1.49
35	1	1.15	0.52	0.87	1.37	0.60	2.57
70	1	1.11	0.43	1.12	1.56	0.67	2.94
35	0	0.84	0.44	1.50	1.36	0.59	2.32
70	0	0.85	0.45	1.94	1.53	0.60	2.65
L.S.D.	5%	0.22	0.07	0.13	0.20	N.S.	0.16
	1%	0.31	N.S.	0.18	0.30	N.S.	0.22

Compost alone did not provide sufficient N for satisfactory sorghum growth during the first year; however, during the second year N availability was improved considerably. Nitrogen uptake by sorghum during the first year was significantly less in plants growing in compost treated soil than where 10-10-10 alone was applied.

This was reversed during the second year which indicated that residual compost was influencing N availability. Nitrogen determinations in oat foliage (Table 6.22) also supports this observation.

TABLE 6.22: EFFECTS OF COMPOST ON N-P-K CONTENTS OF OAT PLANTS

Compost	10-10-10	1968-'69 (%)			1969-'70 (%)		
----Tons/ha--------		N	P	K	N	P	K
0	1	0.66	0.38	1.07	0.65	0.33	1.47
35	1	0.87	0.42	0.97	0.75	0.34	1.37
70	1	0.93	0.47	0.95	0.83	0.32	1.28
35	0	-	-	-	1.07	0.45	1.25
70	0	-	-	-	1.10	0.41	1.24
L.S.D.	5%	0.12	0.03	N.S.	0.13	0.07	N.S.
	1%	0.18	0.04	N.S.	0.19	0.08	N.S.

SUMMARY

Under a strict interpretation of the requirements that an organic material must meet for compost qualification, only the Fairfield material could be so classified. That material had a C/N ratio of 17 as compared to 26 for the Largo compost, 43 for the St. Petersburg compost, and 55 for the Gainesville compost. In addition to a relatively low C/N ratio, the Fairfield compost was pelletized, free of moisture, and agreeable to handle. In contrast, the other composts contained excess moisture and had physical characteristics that made them exceedingly disagreeable to handle. The low soluble salt content (Table 6.1) of the Fairfield compost and the high soluble salt content of the other three composts must also be considered in rating these materials.

The future of municipal refuse composting depends to a large extent on utilization of the finished product. These investigations indicated that the compost was beneficial to the soil and that relatively large amounts could be applied without danger to plants. However, it must be recognized that low N content and high soluble salt content present problems that must be overcome. In both instances the soil microbial population and climatic conditions will alleviate detrimental effects in time, by narrowing the C/N ratio and by leaching soluble salts below the root zone.

USPHS-TVA Composting Project

SUMMARY

The Joint USPHS-TVA Composting Project began operation in June 1967. The purposes were to provide the then Bureau of Solid Waste Management with more comprehensive knowledge about windrow composting as a solid waste management tool and permit better assessment of available information about this subject. Results of investigations and operational experiences from the project from June 1967 to September 1969 are discussed in PB 214,056 by G.E. Stone and C. C. Wiles, directed by C.A. Clemons, from which material in this chapter is derived.

During the period, the plant processed an average of 34 tons of raw refuse per processing day received from a population of 31,200. Approximately 27% of the incoming refuse was rejected as noncompostable and returned to the city's landfill for disposal. The yield of unscreened compost was about 50% of the incoming refuse (wet weight basis) but trouble with the equipment for grinding compost hindered the production of an acceptable grade.

Investigations of the potential hazard to health from pathogenic organisms in compost showed that windrow temperatures of 122° to 130°F maintained for at least seven days destroyed pathogens expected in refuse and those known to be in sewage sludge. Time-temperature studies showed that windrow temperatures at the 1½ ft depth averaged above 140°F for 2 to 3 weeks or more.

Sewage sludge, cow manure, paunch manure, aged poultry manure, animal blood, and pepper canning wastes in varying amounts were all successfully composted with the refuse. In the amounts used they did not greatly affect the composting process or the product. The addition of urea-ammonium nitrate appeared to inhibit microbial activity and resulted in a loss of nitrogen in the product. Although limestone added to the refuse appeared to aid the composting process, it also caused a loss of nitrogen and resulted in a poorer product.

Total construction costs, including costs of modifications made during the period, were $958,375. Actual cost for operating the plant during 1968, at a level less than capacity, was $18.45 per ton of refuse processed. Projected to full capacity, the operating cost in 1969 was $13.40 per ton of refuse processed. Because production of an acceptable grade of compost was delayed, agricultural testing of the compost is only in preliminary stages. Therefore, insufficient data are available to assess the agricultural value of the compost produced.

BACKGROUND

In the early 1960s, F.E. Gartrell of the Tennessee Valley Authority (TVA) proposed a full-scale composting project to be jointly sponsored by TVA, the U.S. Public Health Service (USPHS), and a municipality in the Tennessee Valley. Both USPHS and TVA were interested in a sanitary method of waste disposal and TVA was also interested in the use of compost for soil improvement.

In March 1963, the Division of Agricultural Development, TVA initiated a project called "The Use of Municipal and Industrial Organic Waste in the Production of Soil Amendments and Fertilizers." The compost plant was to investigate the process as a method for disposing of wastes and at the same time reclaiming waste for the benefit of the land.

In August 1964 the USPHS and the TVA agreed upon a joint research and demonstration project for composting solid wastes and sewage sludge. In November 1964 USPHS detailed John S. Wiley to the Division of Health and Safety, (now the Office of Health and Environmental Science) TVA, to collaborate in the detailed planning of the project. Under the agreement, TVA would have financed and constructed the composting plant.

When the Solid Waste Disposal Act was passed in October 1965, the Bureau of Budget directed that the proposed plant be financed from funds available under the terms of the Act to the Department of Health, Education, and Welfare. Johnson City had been selected as the site for the composting project on the basis of two surveys of six Tennessee Valley area cities.

One survey, conducted by TVA engineers and representatives of state and local health departments, covered disposal of refuse and sewage sludge. The other, made by TVA agricultural specialists, investigated the use of chemical and organic fertilizers in the vicinity of these cities. On February 15, 1966, the two agencies and Johnson City, Tennessee, signed a cooperative agreement for the construction and operation of the Joint USPHS-TVA Composting Project, Johnson City, Tennessee. Ground was broken for the plant on May 18, 1966; construction was completed and the plant was put in operation on June 20, 1967.

JOHNSON CITY, TENNESSEE

Johnson City, in Washington County, is in the extreme northeastern corner of

Tennessee. It lies near the junction of the Watauga River and the South Fork Holston River in the headwaters of the Tennessee River system. The terrain immediately surrounding the city ranges from gently rolling on the east and south to very hilly on the west and north. Mountain ranges begin about 5 miles to the southeast and about 20 miles to the west and north, with many peaks rising to 4,000 and some to 6,000 ft above sea level toward the southeast in the Appalachian system. Elevations in the city range from 1,500 to over 1,700 ft.

The Johnson City area does not lie directly within any of the principal storm tracks that cross the country, but comes under the influence of storm centers that pass along the Gulf Coast and then up the Atlantic Coast toward the northeast.

Johnson City, with an area of 15 square miles had an estimated population of 35,300 in 1969. The city and county had an estimated population of 68,500. The trading area had an estimated population of 180,000 and the area within a 50 mile radius had an estimated population of 635,000.

Tobacco is the most important crop of the area. Corn and some small grains are grown in support of the important dairy industry. Raising of beef cattle has increased in the last 20 years to become a significant part of the economy. In the seven counties of the area, there are about 320,600 acres of cropland and 281,800 acres of pasture.

Waste Collection

Residential wastes of 31,200 persons and 40% by weight of the commercial wastes are collected weekly in Johnson City by the Sanitation Department using 5 three man crews with compactor trucks. 60% of the homeowners, however, burn some wastes in their backyards. Others employ private collectors to provide supplemental service once a week. Street sweeping and brush are collected by the Street Department and are deposited in the city's landfill.

Industries in Johnson City generate about 130 tons of solid wastes weekly. The majority of these are hauled to disposal sites by employees of the industries. The Sanitation Department collects from seven industries, totalling about 9 tons per week.

Wastes collected by the Sanitation Department are delivered to a transfer station. Here they are transferred to either a 53 or a 65 cu yd compaction trailer for hauling to the compost plant or the landfill. One load per day of commercial material, about 95% paper, and averaging about 10 tons, is routinely hauled to the landfill.

Hospital wastes and some yarn wastes from an industrial establishment also do not reach the composting plant. When the compaction trailers are out of service for any reason, the 16 cu yd packer trucks used for residential collection deliver their refuse to the composting plant. For the period from January 1968 through June 1969, the compost plant received an average of 33.8 tons of refuse per day for 5 days per week. This collection with the 10 tons per day sent directly to the landfill, is about 2.8 lb per capita per day on a 5 day week basis and 2 lb per capita on the 7 day week basis. The low figure probably does not represent the true per capita generation of refuse in Johnson City.

Composition of the Municipal Refuse

Two studies of the production of refuse per capita and the composition of that refuse have been made in the same residential area of Johnson City. The first was performed by the Division of Technical Operations, Bureau of Solid Waste Management in October 1967, and the second by the staff at Johnson City in July 1968. The first study showed a production rate of 1.1 lb per capita per day of residential waste and the second showed a rate of 1.4 lb per capita per day (Table 7.1).

TABLE 7.1: MUNICIPAL REFUSE PRODUCTION

	October 1967	July 1968
Refuse production (lb/capita/day)	1.1	1.4
Number of homes sampled	136	144
Population sampled	519.8	550.0
Population density (people/home)	3.82*	3.82*
Weight of refuse collected, lb (wet weight basis)	4,003	5,400
Sample period, days	7	7

*Based on 1967 study.

The composition of these samples was found by manually sorting the refuse into ten categories (Table 7.2). This refuse was higher in food wastes and lower in paper wastes than the averages usually reported since it was collected from a residential district.

The refuse received daily at the Johnson City plant probably has a greater paper content. As previously mentioned, one load of commercial waste, with a very high paper content, is not received at the compost plant but is hauled directly to the landfill.

TABLE 7.2: COMPOSITION OF REFUSE

Category	Wet weight, pounds		Percent of total sample	
	October 1967	July 1968	October 1967	July 1968
Paper products	1,820.4	794.5	45.5	34.9
Food wastes	1,036.5	788.5	29.5	34.6
Metals:				
Ferrous		211.5		9.3
Nonferrous		24.5		1.1
Combined	433.0	236.0	10.8	10.4
Glass products	436.1	206.0	10.9	9.0
Plastics	67.0	76.5	1.7	3.4
Leather and rubber	40.0	55.5	1.0	2.4
Yard wastes	63.5	53.0	1.6	2.3
Cloth and synthetics	56.0	47.0	1.3	2.0
Brick, rock, dirt, etc.	39.0	3.5	1.0	0.2
Wood	11.5	18.5	0.3	0.8
Total	4,003.0	2,279.0	100.0	100.0

DESCRIPTION OF THE COMPOSTING PLANT

Equipment and Processing Steps

The composting plant began operating on June 20, 1967 to demonstrate composting of municipal solid wastes by the windrow method. The plant was designed to process during a 5 day week raw refuse and sewage sludge of 33,000 people, approximately the present population of Johnson City. Nominal capacity is 60 tons per day for an 8 hour shift. Figure 7.1 is a schematic layout of the processing steps, and Table 7.3 lists the plant equipment and specifications.

Refuse is delivered to the plant, 5.4 miles distant, in 53 and 65 cu yd compaction trailers from a central transfer station in Johnson City. The refuse is weighed and then dumped into a hopper in the receiving building or onto the paved apron to be later moved into the hopper by a front end loader. Items such as mattresses, bed springs, bicycle frames, wire, which might block or entangle the equipment, are removed. A 6 ft wide plate conveyor, forms the bottom of the hopper and advancing at a rate of 2 to 10 ft per minute, carries the refuse under a vertical leveling gate and drops it onto a 3 ft wide elevating belt.

FIGURE 7.1: SCHEMATIC LAYOUT OF PROCESSING STEPS

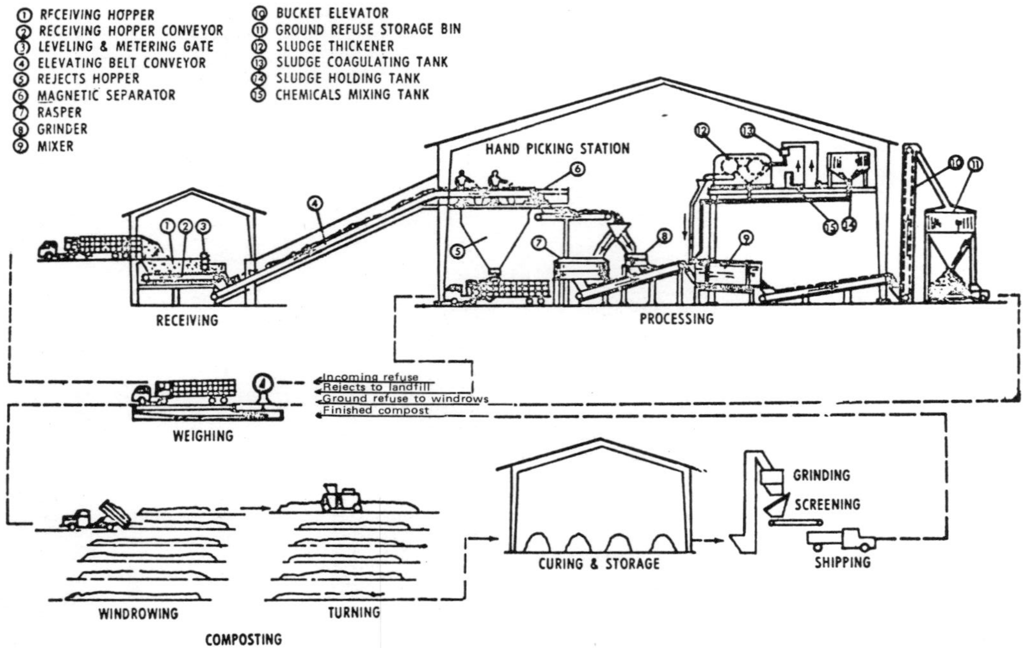

Source: PB 214,056; 1972

TABLE 7.3: DESCRIPTION OF MACHINERY AND EQUIPMENT

Item	Description	Manufacturer & model	Capacity	Power Rating
Truck scale	Mechanical 10 x 34-ft platform scale with tare beam and remote dial weighs incoming refuse.	Fairbanks, Morse, Inc. Model 6507B	30 tons	
Refuse feeder and leveling grate	Double-beaded 5 x 28-ft apron feeder in bottom of hopper (fabricated at plant) moves refuse at 2-10 ft/min; vertical, hydraulically controlled leveling gate regulates flow of refuse.	Websters Mfg., Inc. Feeder and gate driven by Falk helical-geared speed changers, models 51BN2 and 224F2, respectively.	10 tons/hr	5 hp
Raw refuse elevating conveyor	Rubber-covered belt 3-ft wide moves raw refuse at 60 ft/min from receiving building to processing building under cover and then to picking station. Head pulley is permanent magnet (field: 400 Gauss, 5 in. from face at center) for removing ferrous metal.	Continental Conveyor and Equipment Co.	10 tons/hr	2 hp
Raw refuse cross conveyor	Rubber-covered belt 3-ft wide moves refuse at 60 ft/min from picking station to grinders.	Continental Conveyor and Equipment Co.	7.5 tons/hr	1 hp
Rasping machine	Arms on rotating vertical shaft force refuse against rasping pins projecting from sides and bottom of 18-ft-diameter housing until reduced to less than 2-in. particles that drop through perforated steel floor.	Dorr-Oliver N.V., Amsterdam. Type R.T.M. 55 V.S.T.D.	7 tons/hr	40 hp each, 2 motors
Hammermill	Mill with 44 chisel-point swing hammers (21 lb each) reduces refuse to particles that pass grate openings 2 in. by 1.5 to 2.5 in.	Gruendler Crusher and Pulverizer Co.	12 tons/hr	250 hp

(continued)

TABLE 7.3: (continued)

Item	Description	Manufacturer & model	Capacity	Power Rating
Ground refuse cross conveyor	Rubber belt 2.5 ft wide moves ground refuse from hammermill to ground refuse conveyor.	Fabricated by TVA	7.5 tons/hr	1 hp
Ground refuse conveyor	Rubber belt 2 ft wide moves ground refuse from rasper to mixer at 200 ft/min.	Continental Conveyor and Equipment Co.	7.5 tons/hr	1.5 hp
Sludge concentrator	Electric-motor-driven stationary sewage sludge concentrator, traveling screen type, utilizing gravity as the dewatering driving force. The nylon filter cloth moves at a variable speed of 1.5 to 4.5 fpm. The unit is designed to dewater raw sewage sludge containing 3 to 5 percent solids and deliver a sludge cake with approximately 15 percent solids. In addition to the filter, there is a Wallace & Tierman flocculant dosing pump and a Worthington metering pump of the diaphragm type.	Permutit Co. Model DCG-200	Designed for 1200 gal/hr	
Refuse-sludge mixer	Rotary drum mixer, 3 ft in diameter and 10 ft in length. Internal vanes are designed for mixing the refuse and sludge and discharging mixture at outlet.	Designed and fabricated by TVA		
Conveyor for ground refuse-sludge mixture	Moves mixed refuse and sludge from mixer to bucket elevator. Rubber belt type, 2 ft wide.	Continental Conveyor & Equipment Co.	Rated at 890 cu ft/hr or 11.3 tons/hr at belt speed of 200 ft/min	1 hp

(continued)

TABLE 7.3: (continued)

Item	Description	Manufacturer & model	Capacity	Power Rating
Conveyor for ground refuse-sludge mixture	Replaces sludge-ground refuse mixer when mixer is not in use. Belt is 2 ft wide and 15 ft long.*	Fabricated by TVA		1 hp
Conveyor for discharge of ground refuse	Used to transfer ground refuse to trucks when bucket elevator and storage hopper not used. Belt is 2 ft wide and 25 ft long.	Fabricated by TVA		2 hp
Bucket elevator	Vertical centrifugal type. Electric-motor-driven. Buckets are cast malleable iron, 14 x 7 in., 16 in. on center. Elevator used to transfer ground refuse to storage bin.	Fairfield Engineering Co. Model VCE, No. 147	Rated at 1650 cu ft/hr (75% of theoretical) at chain speed of 200 ft/min	3 hp
Storage bin	Cylindrical storage bin designed to hold several truckloads of ground refuse. Discharge to truck through horizontal sliding gate. Bin is 12 ft in diameter at largest dimension.	Designed and fabricated by TVA	About 110 cu yd	
Bucket elevator	Electric-motor-driven, vertical bucket, centrifugal discharge type. Used in moving compost from feed hopper to vibrating screen.	J. B. Ehrsam and Sons Mfg. Co. H-17	Rated at 1715 cu ft/hr (75% of theoretical) at chain speed of 221 ft/min	3 hp
Vibrating screen	Electric-motor-operated. Vibrating inclined screen with interchangeable wire mesh screens of 1/4- and 1/2-in. openings. Used to screen the compost.	Link Belt		

(continued)

TABLE 7.3: (continued)

Item	Description	Manufacturer & model	Capacity	Power Rating
Compost grinding hammermill	Swing hammer type with 30 hammers each weighing approximately 2-3/4 lb. Electric-motor-driven, heavy duty type. Used to regrind compost.	J. B. Sedberry, Inc. 5W-26	10 tons/hr rated, did not grind compost at this rate.	100 hp
Windrow turner	Diesel-engine-driven, self-propelled. A rotating drum turns the windrow as the machine straddles the row. The drum has teeth arranged so as to move the material toward the center of the windrow as it is turned. The material is picked up from the ground, passed over the drum and redeposited behind the machine. The machine moves through the row at approximately 0.27 mph and is capable of turning windrows approximately 8-ft wide and 5-ft high.	General Products of Ohio, Inc. (Cobey) Model 003	500 tons/hr (spec. claim 1500 tons/hr of 700-lb/cu yd material)	100 hp + (122 BHP by specs)
Tractor shovel	Rubber-tired, gasoline-engine-driven front-end loader with general purpose bucket transports refuse, ground refuse, and compost.	International Harvester Company, Model H-50 (Ser. C)	Lifting: 13,200 lb Bucket: Heaped: 3-1/2 cu yd Struck: 3 cu yd	103 bhp at 2,500 RPM

*In late 1969 the mixer was found to be too small for the capacity of the new Gruendler hammermill. As sludge was not being processed, the mixer was temporarily removed and replaced by this conveyor.

Source: PB 214,056; 1972

The belt carries it into the processing building. The receiving building is roofed and enclosed on three sides. The belt conveyor is covered between the receiving and the processing buildings. The processing building is 40 x 60 ft and houses all of the refuse processing operations, the sludge thickener, and the refuse-sludge mixer. The elevating belt conveyor from the receiving building becomes horizontal after it enters the upper level of the processing building. It carries the refuse past a picking station at a rate of 60 ft/min.

Two or more pickers manually remove bulky paper, rags, glass, plastics, metals, and other noncompostable material. The refuse placed in paper or plastic bags may arrive at this point with the bag still intact. Either the pickers must discard the entire bag or permit some cans and bottles to pass since they cannot tear open all the bags. The belt then passes around a magnetic pulley where cans and other ferrous metal objects are separated from the refuse.

Some plants may employ a primary grinding operation prior to magnetic separation to reduce the amount of compostable material that might be trapped by the ferrous materials and removed with them. The rejected material, including that removed at the receiving building, is weighed and hauled 4.6 miles to the Johnson City landfill. About 27% of the incoming material was rejected during the period of this report because no salvaging was practiced.

After removal of noncompostables the refuse proceeds to either a hammermill or a rasping machine for comminution. These are so placed that they can be used alternately for comparison of efficiency and costs of operation and maintenance. The hammermill is of the swing hammer type, having 44 chisel point hammers each weighing 21 lb. The grate bars have 1½ and 2½ openings. The mill is driven at 1,150 rpm by a 250 horsepower electric motor and has a rated capacity of 12 tons/hr.

The rasping machine is similar to that developed by the Dutch NV Vuilafvoer Maatschappij (VAM). It was imported from the Netherlands where it was constructed by the Dorr-Oliver Company. This machine consists of a covered cylindrical housing 18 ft in diameter and 8 ft, 6 inches high not including the height of the conical cover.

The perforated floor is about 3 ft above the bottom of the housing. Revolving arms attached to a vertical shaft in the center scrape the refuse against projecting pins in the floor and on the wall of the housing, finally forcing it through the perforations. The arms are driven by two 40 horsepower electric motors.

Refuse as received has an average moisture content of 39%. When using the rasper, water is added in the receiving hopper to presoften cardboard. Water is also added in the rasper. When the hammermill is used, water is added after grinding at a point ahead of the refuse-sludge mixer. The ground refuse or refuse-sludge mixture is adjusted to a moisture content of 50 to 60% by wet weight. Raw or partially digested sludge containing 3 to 5% dry solids is pumped to the composting plant from the nearby sewage treatment plant. When using sludge from the same population generating the refuse, the water content is greater than that needed to obtain the 50 to 60% moisture content in the ground refuse-sludge mixture to be composted. Thus, especially with an operation using a rasper, sludge dewatering must be provided. For sanitary reasons it is impractical to add sludge to the refuse at any point in the process ahead of those places where the workmen must come into contact with the refuse.

Sludge is thickened in a Permutit Dual Cell Gravity (DCG) Solids Concentrator to a moisture content of about 85 to 88%. This apparatus uses a revolving filter of fine mesh nylon cloth. As a plug of sludge revolves inside the filter it becomes dewatered, rolls up and is discharged. Special coagulants (polyelectrolytes) are used to condition sludge for filtering. The filtrate is returned with other liquid wastes to the sewage treatment plant where it is treated with incoming sewage. The thickened sludge is discharged to a conveyor belt which in turn discharges it to the mixer.

Refuse and thickened sludge are mixed in a horizontal cylindrical mixing drum designed by TVA. As the drum revolves, internal vanes carry the mixture to the discharge end. From the mixer the material is carried to the ground refuse-sludge storage bin from which it is loaded into dump trucks and transported to the windrow area.

The ground refuse-sludge mixture is composted in windrows on a 4.8 acre area graded and stabilized with crushed rock. The windrows are laid down so as to have a section as near to 9 ft wide and 4 to 4½ ft high as possible. They can be as long as 230 ft. The haystack shaped cross-section will shed rain and the mass of a windrow so shaped will contain the heat of decomposition and can be aerated easily. The field is arranged to receive 34 windrows. As the plant operates 5 days a week and one windrow is laid down each working day, each windrow can remain on the field for a period up to 45 days. The active composting time in the field is 35 to 44 days.

During the time the material is being composted on the field it is turned 8 or more times with a special turning machine. This patented machine straddles a windrow, turning it with a rotating toothed drum as it proceeds from one end to the other. Turning aerates the windrows to supply the oxygen needed for aerobic decomposition. To maintain the desirable 50 to 60% moisture content, water can be added as needed before turning. Should rain cause the moisture content to rise above the optimum, the windrows are turned as often as necessary to dispel the excess moisture and return them to the proper wetness.

After the material has composted in the open field for a period of 6 to 7 weeks it is turned, taken from the field, and deposited in the curing and drying shed. This shed is 60 x 200 ft and provides a curing and drying period of 2 weeks, which is frequently insufficient to adequately dry the compost for screening. Following the drying period, compost is screened and reground as required using a vibrating screen and a hammermill for this purpose. The screen is equipped with interchangeable wire mesh screens of ¼ and ½ inch openings. The mill is of the swing hammer type using hammers weighing approximately 2¾ lb each. A 100 horsepower electric motor drives the mill at 3,540 rpm and the rated capacity is 10 tons/hr. Compost is stored either in the unground and unscreened state or in the finished condition in stockpiles.

Construction Costs

Construction of the composting plant began on May 18, 1966, and was completed in June 1967. The total cost, including that of modifications made since startup, is $958,375. Constructed as a plant for both demonstration and research and including some duplicate equipment for comparison, the construction cost is somewhat more than would be required for a municipal plant.

Modifications Since Startup

Many modifications have been made to the plant since operations began. Important ones are tabulated below for the value they may have to future plant designers. They are given in the sequence of accomplishment rather than in the sequence of the process.

Repairing and strengthening the drive of the drum mixer and setting same in level position. Modification of vanes or flights inside the mixing drums. It was found that the original inclination was not needed to carry the material forward. The level setting alleviated a side thrust on the supporting and rotating mechanism.

Addition of wipers to underside of conveyor belts and installation of drip or catch pans under the belts.

Addition of a curb or coaming around the floor which supports the DCG sludge filter and its appurtenances. Spillage in this area had caused cleanup problems.

Laying of plywood flooring over the grating floor at the picking station to prevent spillage to lower level. This also gives more comfort to men who must stand for long periods.

Covering conveyor belt drive motors to protect them from dust and spillage.

Applying thermal tape to pipes and valves in the process building to protect them against freezing. Installing heat lamps at end pulleys of certain conveyor belts to prevent freezing of belt to pulley during shutdown.

Removal of cover on cross-belts from grinders to mixing drum. Covers over belt are not necessary except where they traverse an open area where the wind may cause a problem.

Installing a vertical pipe loop in the sludge pipeline between the metering pump and the sludge-flocculant mixing tank to equalize the hydraulic head of sludge in the holding tank. It was found that the diaphragm sludge metering pump would not operate satisfactorily with the high positive head on the suction side resulting from the elevation of the sludge holding tank.

Removal of strut across receiving hopper at leveling gate end thus removing a cause for jamming. There should be no structural parts which catch refuse over or alongside a hopper.

Modification of chute under the leveling gate. A man must be able to dislodge refuse which may jam the flow at this point.

Installation of dual controls for the leveling gate and the plate conveyor at a station beyond the leveling gate. This gives the man beyond the leveling gate an opportunity to shut down this machinery when necessary.

Erection of targets or poles for aligning and positioning windrows. Ground refuse-sludge mix is deposited by trucks and the drivers need some guide in getting windrows in the correct position.

Adding pads to the bucket of a payloader to give clearance between it and the stabilized rock base of the field to prevent picking up rock with compost when it is being removed or when windrows are being shaped.

Correcting the slippage of the elevating conveyor belt by addition of weights for tension.

Adding skirts to the sides of the elevating conveyor belt to prevent spillage of refuse. This has saved much cleanup time.

Installation of a larger discharge gate for the reject bin. This bin discharges from the bottom. Gates must be as large as possible to avoid arching above the constricted lower part.

Installation of a larger discharge gate for the ground refuse bin. Gates must provide larger openings than those needed for grain, coal, or crushed stone.

Installation of a 12 ton/hr hammermill to replace the original 7½ ton per hour mill.

Backflow preventers were installed in the processing and receiving buildings and the composting field to protect the fresh water system from backflow from the sludge dewatering, washdown operations, and other possible sources of contamination. In the case of the sludge dewatering installation, the water supply piping was redesigned and reworked to enable one backflow preventer, located in a loop of pipe above the highest level of all other lines in the complex, to protect the portable water system from backsiphonage from this operation.

OPERATIONS AND PROCESSES

Production

The plant is designed to process 10 tons/hr or 58 to 60 tons in 6 hr, leaving the remaining 2 hr for cleanup. It has been demonstrated that with refuse of favorable characteristics, the plant can process at this rate. The average capacity however, is about 52 tons/day in 6½ hr of grinding time. The plant has been run as a research project. Although it normally takes all of the refuse delivered by the Sanitation Department of Johnson City, and there have been very few shutdowns, the plant has not been operated to full capacity for extended periods. Processing data for the period January 1968 through June 1969 may be summarized as follows.

The plant received an average of 34 tons/day. Based on the population served of 31,200, this is 2.17 lb per capita per day for the 5 day week and 1.55 lb per capita for a 7 day week. The 10 tons of commercial waste, routinely delivered directly to the landfill, brings the total tonnage collected to 44 tons per day.

This indicates a collection of 2.81 lb per capita per day for the 5 day week and 2.00 lb per capita per day for the 7 day week. This is low compared to the national average. Street sweepings and brush are not included in this quantity. Packing house and produce wastes are sold to piggeries. Much of the industrial waste of the city is hauled privately as is some of the household waste.

Of the incoming material, 27% was rejected as noncompostable and disposed of in the landfill. A seasonal variation in refuse exists, as would be expected. Compost yield from the 11,266 tons of refuse received was about 5,650 tons (30% moisture content by wet weight). All of the compost removed from the field was not weighed.

Receiving

The receiving hopper has a capacity equivalent to two trailer loads of refuse. When the hopper is empty, the 53 and 65 cu yd compaction trailers discharge a small part of their loads into the hopper and then pull away to deposit the remainder on the paved apron in front of the receiving building. A front end loader is used to push refuse from the apron into the hopper. This procedure allows inspection of the refuse, opportunity to remove large items of noncompostables or those which interfere with the movement of the material and breaking up of compacted refuse (thus preventing serious bridging in the hopper). When the hopper contains refuse, the whole trailer load is deposited on the apron to be fed in as needed.

On one occasion, refuse failed to dislodge from the rear door of a 16 cu yd packer truck as it was raised over the receiving hopper. This overbalanced the truck and the front end was lifted upward to almost a vertical position where it was caught by the eaves of the building. Both the truck's hood and the building were damaged. All packer trucks are now required to unload on the apron. Gross picking is necessary at the receiving hopper and at the station behind the leveling gate.

Time records have been kept for the plate conveyor of the hopper and the main elevating conveyor belt which show that this machinery is capable of handling the design capacity of 10 tons/hr or more although the overall average for the last 6 months of 1968 was 9.3 tons/hr. On a day in which 58 tons of incoming refuse were processed, the machinery handled 9.9 tons per hr.

Operating difficulties since startup have been minimal. The drive motor of the plate conveyor required rewiring and a bearing has been rebuilt. Apparently the trouble was not caused by a condition of overloading with refuse. The 328 bolts which fasten the pans of the conveyor to the driving mechanism had to be replaced after some 26 months of operation. It was difficult to keep them tight and the repeated loosening ruined the threads.

The new bolts have a finer thread and are installed with lock washers. An improved program of lubrication has been initiated to reduce heating of bushings, wear of bolts and other conditions.

Due to lack of stiffness the pans or segments of the leveling gate became bowed by the pressure of nonresilient items in the refuse. From time to time the clips attaching the segments to the drive chain broke and had to be rewelded. After processing about 17,000 tons of refuse during 26 months of operation, all of the pans were replaced with pans having heavier bar stiffeners. The elevating conveyor belt has given no trouble since installation of the skirts to prevent spillage. Occasionally cardboard or other material jams under the cover but this is easily removed.

Grinding

The imported rasping machine has performed satisfactorily in the past 18 months of operation. The manufacturer of the rasper states that it has a capacity of 7 tons by wet weight of picked raw refuse per hour. Time and tonnage records kept over a 6 month period show the rasper to be grinding an average of 5.6 tons/hr of refuse from which noncompostables and other objectionable material have been removed.

The rasper must be run for about 10 min as refuse is fed to it before ground material is produced. It then builds up to a maximum production which can be maintained by keeping it full. After the feed belt is stopped, the machine must continue to run at a decreasing rate of production to process the grindable refuse remaining in it. If the rasper is emptied of grindable material at any time during the day, this cycle must be repeated. This characteristic of its operation is reflected in the average tonnage per hour given above.

The character of the refuse has an effect on the rate at which it can be handled. On one day when 61 tons of incoming refuse were processed, the 44 tons of sorted (picked) refuse were ground at a rate of 5.3 tons/hr. On another day, when 59 tons of refuse were processed, the rasper's rate for the 44 tons of picked material was 6.9 tons/hr. On test runs of 1 hr with the rasper full, a rate of 7.1 tons/hr of sorted material received with a moisture content of about 39% was observed.

Water must be added during processing to obtain a moisture content of 60% by wet weight in the material laid down on the field for composting. In practice it is added first in the receiving hopper to soften cardboard and again in the rasper as needed. The rasper is equipped with spray nozzles for this purpose and will not grind dry cardboard without operating difficulties.

The actual weight of material entering the rasper is, therefore, greater than its weight when received at the plant. On the occasion of the rasper's grinding 7.11 tons/hr (39% moisture), it was discharging 10.8 tons of ground refuse with a water content of 60%. At the end of the day there always remains in the rasper some residue. This includes material such as metal cans, items of nonferrous metal, rags, bits of wood, and wire. This material must be removed at intervals either manually or through the rejection opening. When the refuse is difficult to grind, the rasper can become overloaded by the buildup of material. To date this has not caused any mechanical trouble but it does consume time when it is necessary to remove the excess material by hand.

Very little maintenance has been necessary. Not long after startup a bearing in one of the American built motors required repair which was done by the manufacturer. The main bearing of the shaft which carries the rotating arms became overheated soon after operations were begun. This was corrected by a modification of the lubrication of this part.

In March 1969, after 20 months of operation and the grinding of 8,720 tons of sorted refuse, the rasper bottom plates, pin plates, and wear plates were removed and replaced. The 8,720 tons of picked refuse represent nearly 12,000 tons of incoming refuse. In September 1969, after 26 months of operation, two of the rasping arms broke at the point where they are fastened to the rotor. These were rewelded with some steel reinforcement by plant personnel. Three others, not broken, were also strengthened and such trouble is not expected to recur within 2 years.

The hammermill originally installed in the process building did not produce the particle size desired or achieve the specified capacity for grinding 7½ tons/hr of raw refuse. Tests during the first 6 months of plant operation showed a capacity of 5.6 tons/hr. In late 1967 and early 1968, experiments were conducted with several grate size combinations to see if a better particle size could be obtained. The smaller the particle size produced, the greater were the operating difficulties and the smaller the grinding capacity. In March 1968, tests with the original grates showed a production of only 4.2 tons per hr.

On every occasion of operating the mill for grinding refuse, stoppages in the feed chute occurred. Before an arrangement was made whereby the mill could continue to run after the feed belt was stopped, the mill jammed and could not be restarted until it had been opened and cleaned out. After March 8, 1968, the use of the hammermill for grinding of refuse was discontinued.

From March through December 1968, the hammermill was used only for regrinding compost for use by the TVA agriculturist. Between March and October 1968, a total of 624 tons of compost at least 6 weeks old was ground at an average rate of 8.9 tons/hr. The range of production was from 3.7 to 15.6 tons/hr. The dryness and other characteristics of the material and the speed with which it can be fed into the mill affect the rate of grinding. Compost with a water content over 30% by wet weight can give trouble by sticking and jamming. Material that is very dry can create a dust problem.

In mid-October 1968, the hammers were replaced with a set which had been refaced by a buildup of welding rod. A total of 288 tons of compost was subsequently reground at the rate of 13.6 tons/hr. The hammers taken out of the mill in October 1968 had been installed in November 1967, and had been used for 102 hr. Of this time, 27.4 hr had been for grinding refuse. Total weight of refuse for this period of grinding refuse was not determined as it was the accumulation of several short experimental runs. Experience had shown that the hammers of this mill required refacing after 40 hr of refuse grinding. After about 27 hr of grinding, they would have had 13 hr of remaining usefulness for this purpose. Thus, 13 hr for refuse grinding were good for about 74 hr of compost grinding.

At this rate, a set of refaced hammers might serve for nearly 230 hr of compost grinding. This mill had ten 50 lb and four 15 lb hammers operating at 1,800 revolutions per minute, driven by a 125 horsepower motor. The grates were of the bar type. The mill recently installed for regrinding compost is a different type, having narrow hammers weighing about 2¾ lb each, designated to rotate at 3,450 rpm, and powered by a 100 horsepower motor. It is intended for grinding compost that has been rejected by the vibrating screen. Information obtained on the wearing of hammers on the larger mill is not applicable to the new final grinder.

In February 1969, a new and much larger hammermill for refuse grinding was installed. This mill is described in Table 7.3. During the month of March 1969, it had been operated for 47½ hr, grinding 275 tons of refuse, when the hammers were taken out for rebuilding. The hammers were badly worn resulting in less efficient particle size reduction and indicating that removal after approximately 40 hr of operation seems to be necessary. Rebuilding after 30 hr will result in less difficulty in refacing.

The short experience with the new hammermill has shown that the pickers and the magnetic separator may miss large pieces of metal that will give trouble. In one case, a steel ball bearing almost 3 inches in diameter got into the mill. In another case, a large piece of metal was thrown about in the mill with such force that the ½ inch steel cover was bent. A tramp metal trap, which is quite heavy, was installed. This in turn necessitated the installation of a trolley hoist above the hammermill to assist in removing the trap and cover during hammer changes or for other reasons.

Addition of Sewage Sludge

On January 2, 1968, the routine addition of sludge was attempted. Sludge is pumped from the primary sewage treatment plant adjacent to the composting plant into a 2,600 gal holding tank. From this tank it is transferred to a mixing tank by a diaphragm metering pump. An anionic or cationic polyelectrolyte flocculant is introduced into the line ahead of the mixing tank by a positive displacement feed pump. After mixing, the sludge flows into the filter.

In January and February of 1968, subfreezing temperatures on eight occasions made it impossible to process sludge. The processing building was unheated and the freezing of pipelines and even dewatered sludge was experienced. Wrapping of pipelines with thermal tape and the installation of space heat in the building have corrected this problem.

When a breakdown of the sludge filter occurred, there was difficulty in disposing of unneeded sludge due to the arrangement of the sludge return lines. It was decided to add sludge on a 3 day a week basis in order to carry on the essential pathogen survival studies. In July 1968, the plant again attempted to process sludge daily. Of the 103 days on which refuse was processed in the second half of 1968, dewatered sludge was incorporated into ground refuse on 75 days.

Of about 348,000 gal of sludge received from the Johnson City sewage treatment plant, 261,000 gal were processed through the filter. The average number of gallons processed per day of filter operation was 3,480. On 13 days the plant was unable to process sludge because of downtime on the filter. Of the 75 sludge processing days, the filter was operated only part of a day on 12 occasions on account of downtime. Over the 6 month period the added sludge solids averaged about 3.9% of the dry weight of the raw refuse-sludge mix laid down in windrows. On occasion, greater amounts were added for specific experimental windrows.

The DCG filter is capable of dewatering sludge from a water content of 96.9 to 85.0% of wet weight. The capacity (not necessarily raw sludge) is specified as 1,200 gal/hr. The monthly average rate has ranged from 861 to 1,060 gal/hr. The maximum amount of sludge processed on any one day has been 5,643 gal. The equipment installed is not capable of handling all of the sludge generated by the 27,000 population served by the city sewer system.

The flash mixing tank, where the sludge is mixed with a flocculant before it enters the dewatering cells, is equipped with a DC motor, the speed of which is controlled by a rheostat. This rheostat burned out and the mixer was subsequently operated with an AC motor. This operation must be intermittent, however, because the motor runs at too high a speed. Permutit's attempts at obtaining a replacement rheostat have failed as the company that manufactures the speed control unit will not sell parts for it. An alternate solution is to install a gear reducer for the AC motor. Permutit is to supply parts numbers for such a reducer.

The sludge inlet piping to the dewatering cells splits the flow from a flash mixing tank into four parts by using a weir box. Rags and other large materials tend to clog this box, requiring continual attention to insure steady flow. The cause of this trouble lies partly in design, which Permutit intends to correct in future models, and partly in the fact that the Johnson City sewage treatment plant does not have comminution.

Trouble was experienced with the displacement of the sprocket idlers over the cells. When these came loose, the cell moved out of position and became deformed and jammed. The manufacturer recognizes this fault and will design a correction into a test machine being constructed; it is not known if this correction can be applied to the apparatus installed in the composting plant. The nylon mesh filter medium is stretched between rubber seals which in turn are attached to a drive chain which imparts the rotary motion to the cells. The seals were not strong enough to stand the stress of being part of the drive mechanism and still support the weight of the sludge roll between them. Metal inserts in the seals to which the drive chains are bolted proved to be too weak for the purpose. The machine was developed for dewatering digested sludge which forms a smaller roll than the raw sludge treated at the composting project. Raw sludge forms rolls in the order of two to three times the size of those for digested sludge, overloading the seals, and pulling out the inserts. Each time this happened the seal had to be replaced before the operation could be continued, a rather time consuming task.

Due to these troubles, the processing of sludge in the first half of 1969 was intermittent. Permutit has redesigned the seals so that the new inserts have two horizontal projections instead of one, the web is incorporated into the new seal as it is poured instead of being glued on later, and the new seal is poured in one piece instead of several short pieces. The result is a unit construction which should be capable of carrying a much larger load than the original seals.

The manufacturer supplied the project with a full set of these new seals in the late spring of 1969, and one unit of the filter was equipped with them. In July 1969, it was decided to temporarily cease the processing of sludge in order to devote time and funds to other investigations. The new seals have therefore not been tried in continuous operation.

From the above discussion, it can be seen that the sludge operation has fallen below expectations. In defense of the filter, which was still in the developmental stage, there were times when other operations in the plant or difficulties at the sewage treatment plant resulted in a short day's run. There were times when the quality of the sludge as received made it unmanageable and part of it had to be wasted back to the sewage treatment plant. As it is sometimes drawn from the digester and at other times directly from the primary settling basin, the quality of the sludge has been variable. Furthermore, since there is no comminutor at the sewage treatment plant, rags, sticks, etc., may come through and clog the troughs in the DCG. Raw sludge has better dewatering characteristics than digested sludge.

Handling and Storage

The ground refuse-sludge mixture is moved by conveyor from the mixer to a bucket elevator which lifts the material into a 12 ft diameter cylindrical storage bin. The bin was designed to hold several truckloads of material so that a truck could shuttle back and forth from this point to the windrow field. The bin discharged to the truck through a horizontal sliding gate in the bottom. The bottom third of the bin tapers toward this opening.

It was found that due to arching, ground refuse would not flow through the gate after it builds up in the bin. A vertical screw feed was tried, but was unsuccessful. The gate has been enlarged and now a truck can load, make a short trip, and return to be loaded again without trouble. If the truck does not return shortly, however, the bin will clog. The conclusion is that bins should have straight sides and a live floor.

A common ensilage wagon with such a conveyor floor has been tried with success in an experiment.

Windrow Maintenance

Although the windrow turning machine successfully mixes and aerates the composting refuse, structural and mechanical troubles have caused it to have considerable downtime, this being only the second such machine ever built by the manufacturer. The machine as delivered tended to spread the windrows. The addition of teeth to the rotor and changes in their inclination to the center improved the capability to maintain the shape of a windrow. A vexing trouble was the frequent breaking of the drive chain. Overheating of fluid in the hydraulic transmission resulted in failure to transmit sufficient power.

In the spring of 1968, the manufacturer had a mechanical engineer design some improvements and sent a mechanic to the plant to make modifications. Work on the rotor teeth resulted in a machine capable of shaping windrows newly laid down by the dump trucks. A stronger drive chain with an improved idler sprocket arrangement was installed. The hydraulic drive system was improved including cooling of the fluid, some repairs were made to the rotor, and the chassis or frame was somewhat strengthened.

In August 1968, a shaft in the gear train broke on two occasions. Since the motor produces over 100 hp (BHP 122 by specification) and the reducing gear has a nameplate rating of 18 hp, TVA operations staff installed a torque limiter to protect the gear train. Despite the torque limiter, gear train troubles have persisted. Some motor troubles were experienced but these had nothing to do with the design of the turner.

In December, the axle and kingpin of one of the rear (steering) wheels broke. This was expected due to the light construction and the stress put upon it by an unsophisticated steering arrangement. The steering has been improved by the TVA and the stronger kingpins installed.

After the modifications made by the manufacturer in April 1968, the machine moves over the ground at 0.27 miles/hr instead of the 4 miles/hr claimed by the manufacturer. This speed is not appreciably changed in working through a windrow and the machine appears to be able to aerate about 500 tons of compost per hr. Evan at this slow rate, it is capable of turning the 34 windrows in 4 to 5 hr. Rarely would it be necessary to turn all windrows on the same day.

During composting, the windrows tend to slump. To maintain a desired windrow cross-section, they must be reshaped. Before turning, the margin or foot, where the side of the row meets the ground, must be picked up and placed on top or at the end of the windrow to assure that this part of the material will be subjected to the heat generated within the central mass. Windrows must be kept at 50 to 60% moisture content at least until the 28th day. If they become too dry, they must be wetted and turned. Should they become too wet, turning can help to dry them.

In the summer of 1967, wet weather gave some trouble. The summer and fall of 1968, were much drier than were these seasons in 1967. Added to this, Johnson City suffered a severe water shortage and asked users to practice conservation as far as possible. The watering of windrows was cut to a minimum and by the end of September, the use of domestic water for the complete daily washdown of the plant was discontinued. Portable irrigation pipe and a portable pump were brought in from a nearby TVA installation. The pump was set up near the plant on the bank of a creek entering the Watauga River and this supply was used for washing down operations. The city put a new filtration plant into operation in mid 1969 and there is now an ample supply of water.

Curing and Drying

The water content of the digesting refuse is kept between 50 to 60% by wet weight throughout the first 28 days. By the 42nd day, the moisture content is usually between 40 and 50%. Trouble has been experienced in air drying this compost to the 25 to 35% moisture content desired for screening, regrinding, and final disposition. Except in very dry weather, the 2 week storage under cover of the curing shed does not accomplish this moisture reduction. In wet weather, the moisture content can be static for long periods. In dry weather, better results have been obtained by leaving windrows uncovered on the field exposed to the sunlight. Experience indicates that for commercial production of compost in a climate similar to that at Johnson City, mechanical drying equipment may be desirable.

Screening and Grinding

In the summer of 1968, regrinding was done in the hammermill originally installed for grinding raw refuse. No screening was done. In February 1969, a mill procured for regrinding and the vibrating screen were installed. The winter had been wet and extreme difficulty was experienced in obtaining material dry enough to screen or regrind. It was not until late April that compost was dry enough for real testing. None had reached the favorable moisture content of 25%. The use of compost in agriculture is seasonal, in fall and spring. It appears that compost will have to be screened and reground in late summer and stockpiled for fall and spring use.

The compost as taken from the field is still quite hot but has been reduced to an innocuous, earth-like material with no objectionable odor. The composting activity still goes on but at a slower rate. The heat is dissipated slowly from the inner mass of stockpiles, but is reduced. Drying reduces microbial activity and further enhances cooling. Decomposition is not complete, but for many uses, it has reached a practical point in large-scale composting.

Although the vibrating screen and a compost grinding mill were installed in February 1969, operating troubles and dry difficulties have prevented a full evaluation of this phase of the operations. The compost is first screened and that material which does not pass through the screen is ground and then recycled to the screen. Considerable trouble was experienced due to blinding with the

¼ and ½ inch woven wire mesh screens first installed. A perforated plate screen with 7/16 inch openings is now being used. This screen will pass about 50% of fresh compost when the moisture content is between 30 and 35% by weight. Production is 1 to 3 tons/hr, much less than the 10 tons/hr expected.

The hammermill installed for grinding compost has proven inadequate and has experienced excessive hammer wear. A set of hammers furnished with the mill lasted for only 8 hours grinding at the rate of about 2 tons/hr. Because of the excessive cost of replacing or rebuilding the hammers furnished with the mill, hammers fabricated from strap steel are now used and discarded when worn. About 80 tons are ground between changes.

Because of the disappointing performance of the small hammermill, consideration has been given to replacing it with the Williams hammermill formerly used for grinding refuse. This mill has shown a capability of grinding more than 10 tons per hr of compost. A rotary, drum-type screen is being considered to replace or supplement the vibrating screen.

Fly Control

On June 20, 1967, it became evident that flies were a problem. Through the Regional SWP Chief, Region IV, an inspection by an entomologist of the National Communicable Disease Center was arranged. The Division of Health and Safety, TVA, also arranged for a TVA entomologist to visit the plant. It was the opinion of these two entomologists that although flies were breeding on the windrow field, the greater number had come to the plant in the refuse as eggs, larvae, and pupae.

In the several days that it takes for a windrow to reach an elevated temperature, eggs hatch, and larvae migrate to suitable places for pupation. After windrows heated up and were turned twice a week during the first 14 days, breeding was lessened because of the heat destruction of eggs and what breeding appeared occurred mostly at the toe where turning is less efficient and temperature is low.

Before the inspections by the entomologists, the receiving and process buildings had been sprayed with a pyrethrum-piperonyl butoxide space spray and Dibrom in solution had been used on the walls and on the windrows. Application was by a gasoline powered sprayer. Only partial control was obtained.

The CDC entomologist recommended Dimethoate in an emulsifiable concentration for space spraying and as a mist or fog in the windrow area to kill adults. For larvae, Dimethoate wettable powder suspension was recommended for a residual on the walls of the receiving and processing buildings. At the receiving building, larvae could be seen migrating from fresh refuse.

He further recommended that Dimethoate wettable powder be incorporated into the ground refuse before it was laid down in windrows. This material was obtained and tried as a residual and a mist. The equipment on hand did not create a mist of the magnitude necessary and a spray was applied directly to the windrows. Some abatement was accomplished but flies continued to be a problem until cold weather set in. The recommendation for adding the insecticide to the ground refuse was not followed because of the unknown effects upon the flora carrying on the composting and upon the pathogen survival study. The inspecting entomologists pointed out the fact that the once a week collection in Johnson City was related to the large number of larvae in every stage of development found in the incoming refuse. In their opinion, two collections a week would probably aid in the control of flies at the plant.

On July 5, the first significant fly population of the summer of 1968 was observed at the plant. Using a Dynafog Model 40 gph nonthermal fogger or mister and an emulsion of 12 gal of 22.4% Dimethoate emulsifiable concentration in 38 gal of water, control was effected. This dosage is equal to the recommended 6 gal of 50% Dimethoate EC in 50 gal of water. The speed of application was 4 miles/hr and it was found that only the four youngest windrows on the field needed fogging. The dosage was later reduced to 8 gals of concentrate in 42 gal of water with equal control being accomplished.

The receiving building and the apron were sprayed with a Dimethoate emulsion for the residual effect. Although fly larvae could be seen migrating from fresh refuse received at the plant, very few appeared to pupate. On the composting field, there was some breeding of flies in the toe of the windrows. This material is now picked up with a payloader and placed on the top of a windrow or at the end before each turning.

Interviews with county agriculturists and health officials disclosed that fly populations had been abnormally high on the refuse dumps in nearby Elizabethton and Jonesboro and in dairy barns in the summer of 1967, in contrast to abnormally low populations in the summer of 1968. The rainfall for the area in the period July through September in 1967, was 11.14 inches (normal being 11.97 inches) with 19 consecutive days of rain in July. In 1968, the precipitation for the area was 6.14 inches with a comparatively dry July. The experience of 1968, therefore, did not show to what extent the controls practiced would have been effective in 1967.

In the period July through September in 1969, the rainfall in the area was 14.60 inches, as compared to the normal 11.97 with 8.18 inches in July mostly occurring as thundershowers followed by sunlight. At the plant, where a rain gauge was being maintained, the rainfall was 9.65 inches, with the July precipitation well below that for the area. At no time during that summer did flies become a problem, indicating that the control measures described above appear efficient.

Removal of Plastics and Glass

Plastics, particularly plastic film, present a problem. Shreds of plastic film persist through composting and are carried from the windrows by the wind, causing an unsightly appearance of the grounds. Regrinding and screening reduce the size and amount of the shreds to where they are unobjectionable in the finished product.

As it would be advantageous to remove plastic film before the refuse-sludge mix goes onto the field, TVA has conducted experiments for film removal. Several arrangements of apparatus have been tried, using the principle of separating the plastic film from the ground refuse solely by air currents. Although these showed promise, a significant amount of film has not been removed in this way.

Glass particles are objectionable if they are not ground fine enough or if sharp pieces remain after grinding. At Johnson City, when a rasper is used, glass is not reduced to fine particles as it is in the two or more stages of hammermill grinding used in most other plants. Trials of a drum-type ballistic separator did not show sufficient removal of glass.

TVA also unsuccessfully experimented with an apparatus consisting of a rotating cone on a vertical axis, mounted inside a cylindrical tank. The theory of operating this apparatus was that the refuse falling on the cone, which had an apex angle of about 152°, would be slung off. An attempt would then be made to collect the glass and other dense particles which bounced off the wall of the tank at a trajectory different from that of the organic material. Beneath the cone, suction was to be used to remove plastic film loosened from the stream of refuse by the motion imparted by the cone.

COMPOSTING PROCESS STUDIES

Temperature Studies

In the first year of operation a considerable body of information was accumulated on temperatures attained in the composting refuse. Readings were taken with a portable single probe thermistor type telethermometer. Observations were made on 152 windrows at the 1½ ft depth over the period June 20, 1967 to June 14, 1968. Temperatures over 140°F were recorded in all series for a number of composting days.

For a series of windrows which spanned a period of very cold weather in January and February 1968, a shorter period of high temperatures was observed than for the other series of observations. Figure 7.2 shows results of observations on 6 selected windrows over the period of March 6 to June 14, 1968.

In the period April 26 through June 26, 1968, continuous temperature observations in a windrow were made with a YSI 12 point thermistor temperature recorder. The recorder monitored each point for 5 min at 10 points within the windrow, one point for the ambient temperature, and a point for a register mark.

The instrument cycled the 12 channels once every hour, recording about 11,000 points for the 42 day composting period. Temperatures were recorded at depths of 5, 13, 23, 32, and 41 inches at two stations 10 ft apart. The windrow height was 45 inches. The windrow was turned just before inserting the thermistor probes and again on the 4th, 7th, and 11th day.

FIGURE 7.2: AVERAGE TEMPERATURES OF SIX SELECTED WINDROWS

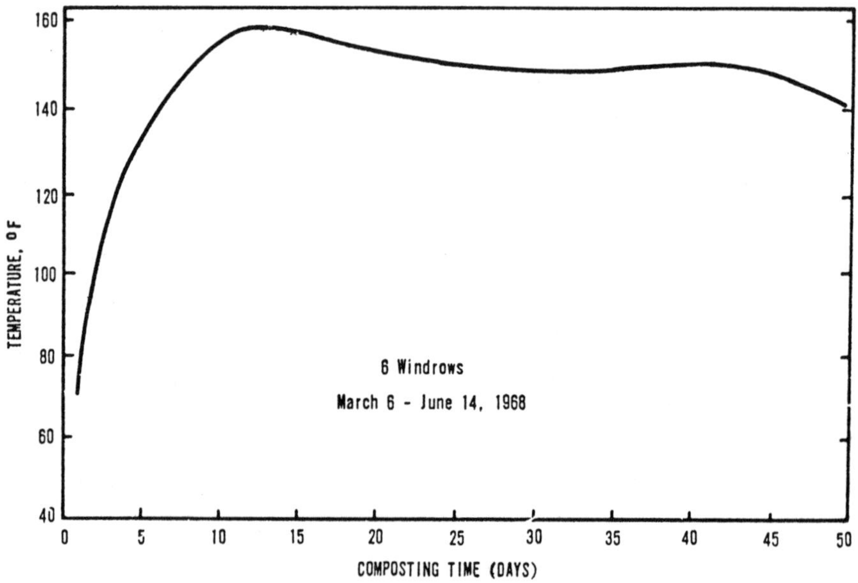

Source: PB 214,056; 1972

Immediately after a turning the windrow has a uniform temperature, after which the various depths reach different temperatures. This results in three groups of data over the period illustrated. Figure 7.3 shows the curve for the 22 and 32 inch depths at Station 1 over 42 days. Station 2 exhibited a similar pattern. Breaks in the curve show points of turnings.

Interior temperatures change rapidly during the first 2 weeks of composting after which the interior temperatures show no great fluctuations. Surface temperatures, however, vary considerably during the entire composting period, depending upon season and weather conditions.

In the period November 4 through December 30, 1968, temperatures were continuously recorded for another windrow at two stations. Figure 7.4 shows the curve constructed from the averaged readings at middepth at the two stations over 42 days of composting. Ambient temperatures were low, reaching 5°F.

Figure 7.5 shows the relationship of position to temperature at various locations around this windrow at a depth of 8 inches and at middepth in the period of 31 to 35 days of composting. The higher temperatures in the upper parts of a windrow and the difference for the south and north sides are to be noted.

FIGURE 7.3: TEMPERATURE PROFILE OF WINDROW 17E

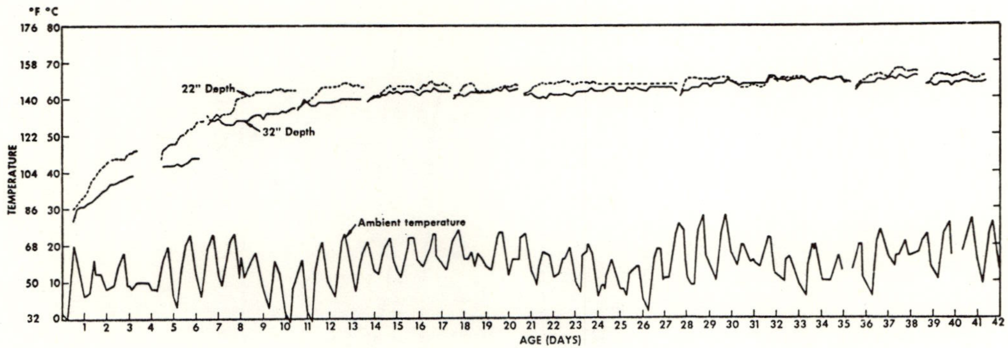

FIGURE 7.4: TEMPERATURE PROFILE OF WINDROW 17H

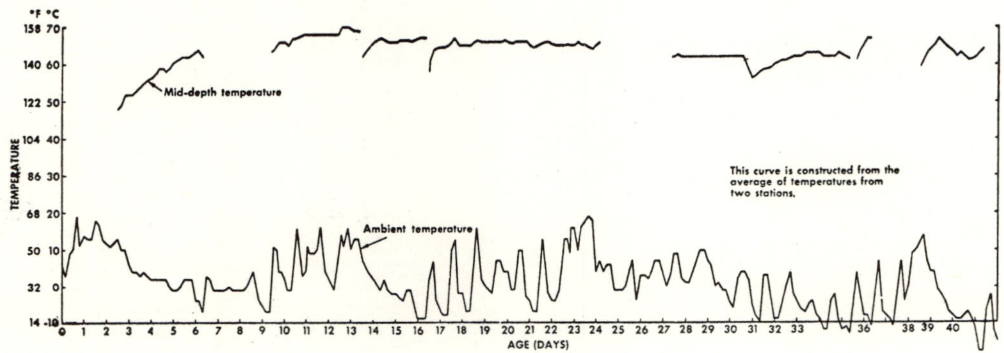

Source: PB 214,056; 1972

FIGURE 7.5: TEMPERATURE PROFILES AT THE 8 INCH DEPTH

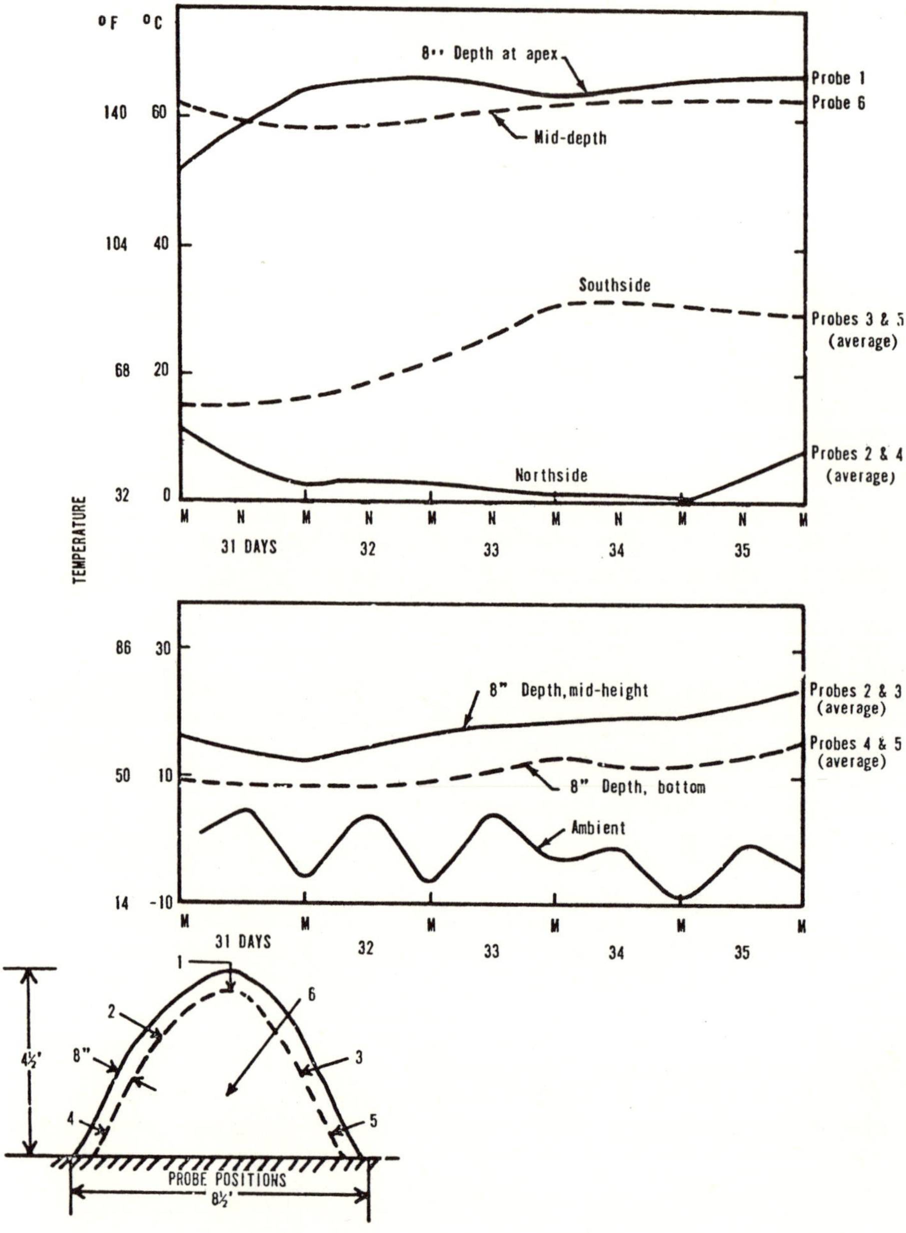

Source: PB 214,056; 1972

Conclusions to be drawn from these detailed observations of temperature are shown below.

- Temperatures vary with depth in all windrows. Temperatures above 140°F (60°C) are reached and maintained for significant periods of time in the inner mass of windrows.
- These variations are more pronounced early in the composting cycle.
- Immediately after turning, the temperature at all channels were found to be within 3.6°F (2°C) of one another.
- After a turning, windrows quickly return to the temperature existing before the turning (Figure 7.3).
- Excepting the outside 8 inches of material, weather conditions appear to have little effect on the windrow temperatures.
- The surface temperatures at the apex of the windrows were often observed to be as high or sometimes higher than the middepth temperature.
- The temperature on the outside 8 inches of the windrow decreased from the apex to the ground. This is considered due to convection currents rising from the bottom to the top inside the windrow. It is interesting to note that these may aid in the aeration of the compost.
- The surface temperatures on the north or shady side of windrow 17H were observed to be lower than those on the south or sunny side.

Effect of Windrow Turning Frequency

Ten windrows were studied for the effect of various turning schedules. The number of turnings during the 49 day cycle for these windrows was 0 to 14. Temperatures were taken once a week throughout the test period. The relative degree of decomposition attained was judged on the basis of appearance, odor, and the amount of carbohydrate reduction.

The present windrow turning schedule is twice a week for the first 3 weeks and once a week thereafter. This routine was based on the higher temperature pattern reached with this schedule, compared with the greater cost of turning more often with lower temperatures resulting. The minimum number of turnings which will result in satisfactory compost is once a week for the entire cycle.

It was found that the best decomposition in the 49 days was obtained by turning the windrow twice each week throughout the cycle. This windrow did not show as good a temperature pattern as the one turned on the present schedule (which attained the highest overall temperatures). It did, however, attain temperatures higher than any windrows turned fewer times.

The windrow on the present schedule showed the next best decomposition and attained the highest overall temperatures. Other windrows, turned less frequently, showed decreasing temperatures. After turning these windrows, however, the temperatures rose 20° to 30°F within 5 days and then declined.

The temperature in the windrow not turned rose to about 130°F in 14 days, then dropped steadily throughout the remainder of the test.

Effect of Composting Sewage Sludge with Refuse

Assuming that (a) a population generates refuse at the rate of 5 lb/day/capita (a figure near the national average) and that it is received at the compost plant containing 35% moisture by wet weight, and (b) that raw sewage sludge solids of the same population are generated at 54 grams of 0.119 lb/capita/day, dry weight, and (c) that rejected noncompostables will amount to 26% of the incoming refuse, then the percentage of dry compostable material (sludge and picked refuse) represented by the dry sewage sludge solids is:

$$100 \times \frac{0.119}{0.119 + (5 \times 0.65 \times 0.74)} = 4.7\%$$

or about 5%. This is the proportion of sludge to refuse to be normally expected from the same population producing the refuse. At Johnson City, although the per capita production of refuse is lower than average, the available sludge handling equipment limits the amount added to refuse to between 3 and 5% of the total dry weight of the refuse-sludge mixture for normal operations. Obtaining a uniform sludge content has not been possible.

Using temperature as a parameter of composting activity, three pairs of windrows were closely observed. In each case a pair consisted of one windrow containing raw sludge and one of refuse alone. The temperature profiles for pairs having windrows containing 2, 3 to 5, and 9% sludge solids were obtained, Figure 7.6 showing that for 3 to 5% sludge solids.

Temperatures were taken at a depth of 1½ ft. The conclusion drawn is that the amounts of sludge being processed, and even up to 9% sludge solids, had no significant effect on the temperatures reached in composting.

The pH profiles of a pair of windrows, one of which contained 2% of raw sludge solids by dry weight showed no significant effect of sludge addition. Figure 7.7 shows the average pH values at various ages for a number of windrows with 3 to 5% raw sludge solids and a number of windrows without sludge. The pH profiles for a pair of windrows, one of which contained 9% raw sludge solids showed a difference in the pH of the masses, the value being higher in the earlier stages in the windrow containing sludge.

Assuming 0.85% nitrogen in fresh ground refuse and 3% in sludge solids, the use of sludge in the amount of 5% of the sludge-refuse mixture (dry solids basis) would result in a finished compost containing 1.05% nitrogen. This calculation assumes that the loss in nitrogen and the loss in weight from sludge during decomposition is similar to the loss from refuse. The addition of sludge in the quantities used does not significantly add to the quality of finished compost as gauged by the nitrogen content. At Johnson City, actual observations show the compost made of refuse and sludge averages 1% by dry weight of nitrogen content.

FIGURE 7.6: TEMPERATURES IN WINDROWS WITH 3 TO 5% SLUDGE SOLIDS

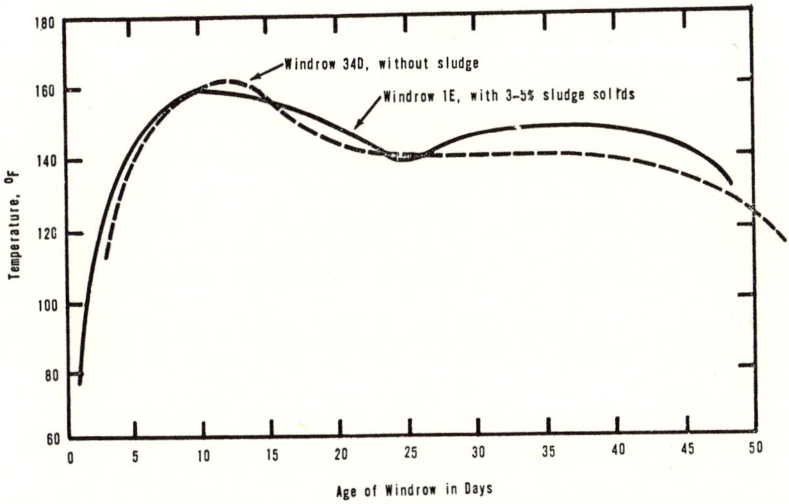

FIGURE 7.7: AVERAGE pH OF WINDROWS WITH 3 TO 5% RAW SLUDGE SOLIDS

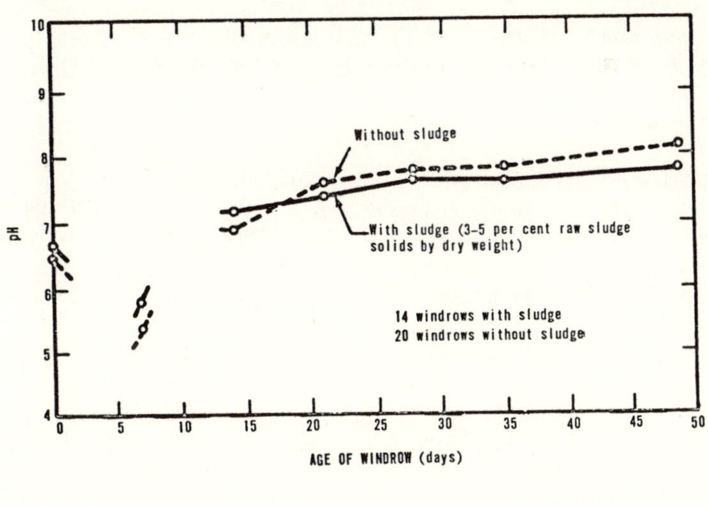

Source: PB 214,056; 1972

Although the carbon to nitrogen ratios of refuse and compost are discussed under a subsequent heading, the observed effect of sludge solids on this important relationship should be mentioned here. The initial C/N ratios of a group of seven windrows containing 3 to 5% sludge solids showed an average of 44.4. At 42 days of composting a similar group of five windrows showed an average of 33.9 and at 49 days two of these showed an average of 31.9. A special test windrow prepared without sludge showed an initial ratio of 44.4 and at 49 days a ratio of 33.0.

The observations of the rate of heating, the pH profile, and the C/N ratios lead to the conclusion that the addition of sludge in the amounts which would normally be available has no appreciable effect on the composting process for refuse similar to that collected in Johnson City. For refuse with a low organic content, the effect of the addition of this amount of sludge may be greater.

It is concluded that sewage sludge solids can be successfully composted with municipal refuse. In the amounts normally available, the proportion of sewage solids to refuse will not greatly affect the rate of decomposition or the quality of the finished compost as measured by nitrogen content. Where greater amounts are available, the nitrogen content and appearance of the compost can be improved.

Effect of Adding a Nitrogen Compound

Under the supervision of the TVA agriculturist, urea-ammonium nitrate containing about 27% nitrogen was incorporated into composting refuse in several tests. The windrows were kept under observation for temperature as a parameter of composting activity and the chemist performed the chemical tests and the moisture determinations. The observed effects are discussed below.

Test 1: Enough urea-ammonia nitrate was added to a section of windrow 8G on the first day to bring the nitrogen content from 0.94 to 3.5% by dry weight of the mass. This windrow contained raw sewage sludge and had a moisture content of 62.1% by wet weight.

The second section was left as a control. It was immediately noted that the pH of the test portion was 8.2 against the normal 6.6 in the control. The pH did not thereafter drop to the low normal obtained in the control. At the 11th day, when the control was at pH 5.00, the test row showed a pH of 7.55. By the 14th day the test row pH was over 8 while the control was barely over 7.

After 21 days the two remained close together with a pH slightly over 8. This change in pH appeared to be linked with the microbiological activity as the temperature in the test row lagged well behind and never reached that of the control. Decomposition was slowed and although the test windrow contained more nitrogen (2.75%) than did the control (1.14%) at 42 days, there had been a significant loss.

Test 2: In this test the urea-ammonium nitrate was added on the 21st day of composting in two windrows, 11M and 12M. Neither of these contained sewage sludge. In 11M, where the nitrogen content was raised to 2.35% by dry weight, the pH was immediately elevated and the temperature of the windrow dropped radically. By the 26th day the temperature had risen to about 150°F and was about normal thereafter. The final nitrogen content was 1.41%, showing a loss.

In windrow 12M, the nitrogen content was brought up to 4.63% by dry weight. The pH reacted similarly to that of 11M but the temperature fell and remained low. The final nitrogen content was 3.35%, showing a loss. It is concluded from these tests that the addition of urea-ammonium nitrate in the amounts used appear to inhibit microbiological activity and result in a loss of nitrogen.

Effect of Adding a Buffering Agent

The pH of fresh refuse or refuse containing normal amounts of sewage sludge is about 6 and may drop to between 4 and 5 soon after being laid down for composting. Often in farm and gardening practice crushed oyster shells, crushed limestone, or even lime is added to raise the pH at the start to accelerate decomposition. At Johnson City, three trials were run using limestone.

In the first trial, 3/16 inch crushed limestone was added in the amount of 21.2% of the total dry weight of the refuse-sludge-limestone mixture. The pH was raised initially to 7.5 but fell to 6 by the 16th day, after which it rose slowly to 8.5 on the 42nd day. The control was also atypical with respect to pH but exhibited the normal immediate drop from an initial 6 to 5.2 in 3 days after which it slowly rose to 8.5 on the 42nd day.

The temperature of the test windrow rose more sharply than did that of the control but dropped after 10 days due to a deficiency of moisture. Water was added and the temperature rose again but not as quickly as would have been expected.

Limestone dust was added at the rate of 9 and 16% by dry weight to two portions of a test windrow prepared from fresh refuse only. The effect in each case was to raise the initial pH and both test sections showed an atypical rise and dip in pH in the first week. Again the attainment of high temperature was accelerated.

Analysis for nitrogen gave the results shown in Table 7.4. Although the temperature rise was accelerated, there was a considerable loss of nitrogen which resulted in a poorer compost.

The results of these field tests are similar to those of the laboratory tests conducted by the University of California at Berkeley.

TABLE 7.4: EFFECT OF BUFFERING AGENT NITROGEN CONTENT

Windrow	Additives	Nitrogen content by % of dry weight 0 day	21st day	49th day
Test No. 1				
Control	3 to 5% sludge solids	0.81		1.02
Test	Same as above with 21.2% 3/16" crushed limestone	0.74		0.57
Test No. 2				
Control	None	0.84	0.82	0.96
Test	9% limestone dust	0.73	0.76	0.66
Test	16% limestone dust	0.64	0.61	0.49

Effect of Composting Other Wastes with Refuse

The TVA agriculturist obtained quantities of cow manure, paunch manure, aged poultry (chicken) manure, animal blood, and pepper canning wastes for incorporation into composting refuse to investigate the possibilities of this method of disposal of such wastes and to see if the finished compost was improved by addition of them. As with the windrows to which urea-ammonium nitrate was added, temperature was used as the parameter of composting activity. Tests with the several wastes are discussed as follows.

Cow Manure: Cow manure, in an amount making it 15.4% of the mixture of manure and refuse by dry weight, was added on day 1 to a section of windrow 17N. No sewage sludge was incorporated in this windrow. The initial nitrogen content of the mixture by dry weight was 1.12% and of the control 0.80%. Final nitrogen content for the composted mixture was 1.14% as against 0.93% for the control. Temperature was slow to rise, reaching in 4 weeks what had been reached in 2 weeks by the control (154°F).

The pH of the mixture initially was lower than that of the control, 4.4 to 6.2. The pH of the control dipped then rose characteristically to nearly 8 in 2 weeks at which time the mixture also had reached the same value. Decomposition of the mixture in this proportion, as indicated by temperatures, appeared to be slowed and the mixture would have to remain on the field longer than refuse alone.

Paunch Manure: Paunch manure from a local slaughterhouse, in an amount equal to 14.9% of the mixture by dry weight, was added on day 1 to a portion of windrow 13G. This windrow contained between 2 and 5% raw sewage sludge by dry weight. The initial nitrogen content of the mixture was 1.22% against 0.97% for the control. The paunch manure itself contained 2.48% nitrogen by dry weight. The temperature curves for the 42 days of composting were identical for the test and control portions. The 42 day nitrogen of the test row was 1.26% and of the control 1.04%.

Chicken Manure: Chicken manure, in an amount equal to 21.2% of the mixture by dry weight, was added on day 1 to a portion of windrow 19G. The chicken manure had been made available to the project on the occasion of a cleanout of droppings from a local farm. A portion of it had undergone decomposition for as much as 6 months. Its nitrogen content was 2.40%, similar to that of paunch manure, and the initial nitrogen content of the mixture was 1.37% and of the control 0.84%. The raw refuse contained between 2 and 5% sludge solids.

Although the initial nitrogen content of the two portions was not greatly different from those of the paunch manure test, the test portion reached its maximum temperature more quickly than did the control. On day 16 the test portion had reached 165°F while the control showed only 150°F. In this experiment the control did not act normally but the chicken manure hastened the activity in this particular batch of compost. The 42 day nitrogen content of the test portion was 1.25% against the 1.05% for the control.

The experiment was repeated with fresh chicken manure in windrow 15N with a concentration of 20.1%. This windrow contained no sewage sludge. Initial nitrogen content was 1.19% for the mixture and 0.73% for the control. Final nitrogen contents were 1.32% for the experimental portion and 0.99% for the control. The temperature of the portion containing the chicken manure approximately followed that of the control to about the third week when the profile for the test row fell below.

By 4½ weeks the test portion was about 12°F below that of the control. It then began to pick up and at the end of 6 weeks was again approximately that of the control. The pH of the mixture reached 8.8 when the control had reached 8. At 6 weeks the pH of the mixture was still somewhat higher than that of the control. The fresh manure and the composting mixture released large amounts of ammonia which accounts for the high pH and the lower temperatures observed.

Beef Blood: Beef blood from a small slaughtering establishment was added on day 2 to a portion of a windrow containing about 3% sewage sludge solids. The 40 gal of blood, weighing 354 lb, contained about 110 lb of solids. The mixture of refuse, sludge, and blood contained about 1.3% blood solids by dry weight. Blood is high in nitrogen, 10 to 14% by dry weight, and this rather small addition was estimated to raise the nitrogen content of the windrow from 0.91 to about 1.0%.

The addition of blood appeared to retard heating, as compared to the control portion, in the first 25 to 27 days. After that, the test portion exhibited temperatures exceeding those of the control. The 42 day nitrogen content for the test portion was 1.25% versus the 1.05% of the control. This experiment showed that slaughterhouse blood can be composted with refuse without trouble and may have some value in enriching the compost. The experiment was not repeated due to the difficulty in obtaining blood.

Pepper Canning Wastes: A batch of the wastes of a pimento pepper canning factory, consisting of the pepper cores, rejected peppers, and skins, was incorporated into a portion of a windrow of refuse containing about 3% sludge solids on day 3.

The pepper waste represented 14% of the total dry weight of the mix. The pepper wastes contained 2.82% nitrogen on a dry basis and the initial nitrogen content of the mix was about 1.1%. The temperature profiles of the test and control portions of the windrow were identical during the composting period. The final nitrogen content of the test portion was 1.3%.

Effect of Covering a Windrow with Plastic Sheeting

Half the length of a windrow was covered with plastic sheeting in April 1968 and the other half was left uncovered. The plastic was removed for turning and replaced immediately after turning. Temperatures were taken of both parts in the center and near the surface at a depth of 4 inches. Some of these data are listed in Table 7.5.

TABLE 7.5: EFFECT OF PLASTIC COVER ON WINDROW TEMPERATURE

Age of windrow in days	Covered		Uncovered	
	Center	At 4" depth	Center	At 4" depth
0	65°	65°	65°	65°
7	116°	114°	144°	139°
14	125°	124°	145°	120°
21	135°	133°	150°	112°
28	137°	134°	149°	131°
35	143°	140°	152°	135°

As can be seen from this tabulation, in the covered portion of the windrow the temperatures in the center of the mass lagged substantially behind those in the uncovered portion. However, after 7 days the surface temperatures were considerably higher in the covered portion due to the greenhouse effect of the plastic. The pH of both parts was about the same. The odors of both parts after turning were about the same, the part under the plastic not being offensive.

Effect of Covering Windrows with Old Compost

In cold weather a retardation of heating has been observed in the windrows. To evaluate the usefulness of covering the new windrows with old compost, three windrows were thus treated.

Windrow 14J was covered with a 12 inch layer of old compost for half its length on zero day. The covered portion showed a steeper rise in temperature at the interface between the old and new than did the control near the surface. It reached 106°F in 3 days, while the control required 4 days to reach this temperature.

At 5 days, the control had reached 116°F against the 108°F of the covered portion. On the seventh day, the control was at 112°F and the covered portion at 108°F. Thereafter they both cooled, the control to 107°F and the covered portion to 98°F, by the ninth day, showing need for a turning.

In the center of the mass, windrow 14J reached a higher temperature, 108°F, than the control by the third day. It thereafter became cooler. The center of the control reached a temperature of 110°F in 5 days, 142°F at 9 days, and then began to cool for lack of aeration.

Windrow 16J was covered similarly on the third day. On that day, immediately after covering, the test portion showed 56°F and the control 48°F at the interface and surface, respectively. On the fifth day, the uncovered portion had reached 118°F against the test portion's 103°F. At 10 days, the control was at 102°F against the test portion reading of 104°F.

The center of the mass of uncovered portion of 16J was 13°F lower than that of the test portion on the day they were covered. By the fifth day, the test portion had reached 95°F against the 114°F of the control. By the seventh day, the test portion was only 92°F at its center while the control was at 135°F. Both cooled thereafter for need of a turning.

Windrow 13J was similarly covered on the seventh day. Starting at 4° apart on that day, they both had reached about 120°F on the next day at the interface. From there on, the covered portion showed an increase in temperature at the interface to the fifteenth day, when it was at 131°F. The control showed only 109°F.

In the inner mass, the control steadily rose from 116° to 151°F on the thirteenth day while the test portion had declined from 114° to 104°F. The ambient temperature during this period was about 32°F. All rows were turned before being covered.

It is apparent from these studies that covering windrows at these times has an adverse effect on the temperatures in the center of the windrow, with no significant advantage being gained with surface temperatures. This was observed when they were covered with plastic, and is apparently due to preventing the windrow from breathing although the effect of the cover in compacting the underlying refuse may also be a factor.

Extrapolating from the results of the windrow covered at 7 days, it appears that the time to cover a windrow to achieve thermal kill at the surface is after it has reached its maximum temperature (two to three weeks).

Such a procedure is under consideration but will require investigation.

pH Observations

Observations of pH changes with age revealed the characteristic curve as shown in Figure 7.8. The pH showed an acid condition for the first week generally with some values below 5.0. As the pH rose above 6.0 to a range of 7.5 to 8.0 during the second week, the temperatures also increased. The pH leveled off in a range of 7.8 to 8.5 for the rest of the composting and curing time.

FIGURE 7.8: AVERAGE pH IN WINDROWS AT 1½ FOOT DEPTH

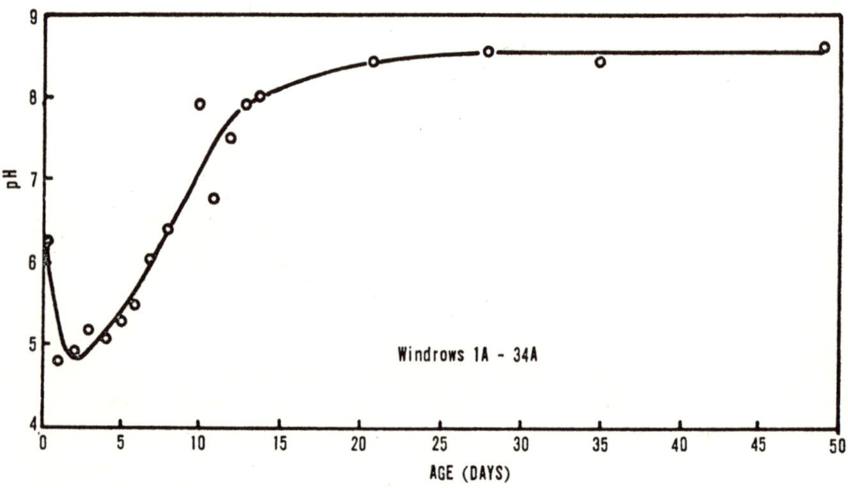

Source: PB 214,056; 1972

MICROBIOLOGICAL STUDIES

Survival of *Mycobacterium phlei*

The Staff Microbiologist conducted studies of the survival in composting of *Mycobactorium phlei* Strain 41 obtained from Dr. George Kubica, National Communicable Disease Center, Atlanta, Georgia. This strain, which is rather thermophilic, was used rather than *Mycobacterium tuberculosis.* A nonvirulent strain of the latter was concurrently being used in survival studies by another investigator. During preliminary experiments in the project laboratory, attempts at the isolation of *Mycobacterium phlei* from seeded raw ground refuse samples by classical published methods proved unsuccessful.

Therefore, all insertions were of necessity made with cultures grown on Lowenstein-Jensen (L-J) agar slants at 45°C for 3 days. All insertions were prepared in duplicate and inserted on either zero day or day 14 at depths of 2 inches, middepth, and in the toe of the windrow. At selected time intervals, the duplicate slants were removed from the windrows.

Subcultures were made by washing each slant with 0.5 ml of sterile phosphate buffer. The buffer-cell suspension was then transferred to a new L-J slant and incubated for at least 10 days at 37°C. Viable cultures would usually produce detectable growth within 3 days. A total of 136 sets of duplicate sample (224 tubes) was inserted into 12 windrows.

No samples inserted at middepth on zero day or day 1 were found to contain viable cells after 14 days of composting. For the insertions at the depth of 2 inches, viable cells were found after 49 days where the temperature reached only 92°F. In other windrows none were found at compost ages over 21 days where temperatures subsequently reached 118°F.

Where temperatures reached 128°F or over at the 2 inch depth, no viable cells were found. This condition was usually obtained by the 14th day. For samples inserted in the toe of the windrow, viable cells were found to the point at which temperatures reached 134° to 136°F. In one case, viable cells were found at middepth at a temperature of 138°F on the seventh day of composting. Samples retrieved at greater ages were not viable.

From the data obtained in these studies it appears that temperatures as low as 128°F are sufficient to kill *Mycobacterium phlei* provided the time of exposure is sufficient. Seven days at 128°F should be sufficient with shorter times for temperatures over 130°F.

Insertions were made under artificial conditions due to the fact that they were slant cultures and not directly exposed to the compost. Temperature, therefore, was the sole factor in their destruction. Where proper mixing is practiced and temperatures of 140°F or over are obtained, *Mycobacterium phlei* would be destroyed.

Pathogen Survival Studies

The intensive studies on the survival of pathogens and parasites in the windrow composting of refuse and sewage sludge were carried out under two contracts (PH-86-67-112 and PH-86-68-143) with East Tennessee State University in the period July 1967 through June 1969.

The first contract studied the occurrence and survival of pathogens and parasites. The second covered the actual insertion of organisms with the compost. The findings of the two studies have not yet been published by Dr. William L. Gaby, under whose supervision they were made.

The organisms for which the search for occurrence was made included the following:

- Coliforms
- Fecal coliforms
- Fecal streptococci
- Coagulase positive staphylococci
- *Salmonella* species
- *Shigella* species
- Enteroviruses (such as polio)
- Pathogenic fungi
- Parasitic organisms (protozoa, cestodes, and nematodes)

In this phase, 602 samples were taken from 30 windrows and from the fresh refuse, sewage sludge, and fresh sludge-refuse mix on the days those particular windrows were laid down. The sludge was raw or in a partially digested state.

The duration of the composting process at Johnson City is 49 to 56 days, 35 to 42 days of which are on the composting field and the remaining 2 weeks in the curing stage, either in the open or under a shed. Samples were taken from windrows at various intervals during the process and on the terminal day. The samples also were taken from several positions within the windrows.

The studies showed that there was a consistent, inverse relationship between the number of total and fecal coliforms in compost and the windrow temperature. Temperatures of 120° to 130°F (49° to 55°C) were sufficient to reduce coliform populations significantly, often to levels lower than the minimal level of detection by the Most Probable Numbers Method.

However, the temperature decrease occurring at the latter stages of the composting process allowed reestablishment of significant numbers (10^2 to 10^5/gram) of coliforms. Fecal streptococci did not appear to be as heat sensitive and maintained populations as high as 10^6/gram even when temperatures reached 130° to 140°F (55° to 60°C).

Salmonella species were frequently isolated from the raw sewage sludge; however, no *Salmonella* or *Shigella* species were isolated from samples of windrows over 7 days old. Coagulase positive staphylococci were isolated only from raw refuse, never from sludge, and found in only one windrow on the 49th day.

All other samples were negative for coagulase-positive staphylococci after the first day on the field. The studies under the ETSU contract did not give conclusive results for pathogenic fungi. No enteroviruses were ever isolated from sewage sludge, raw refuse, or any compost sample at any time.

In the parasite detection studies, 3% of the total number of samples (total of 602) of fresh to 49 day compost were positive for one or more parasitic ova or cysts. Of the 49 day samples taken, 8 of 135 (6%) were positive for parasites. It is of importance here that these positive findings are based on the identification of parasitic forms which were morphologically intact and not upon actual viability tests. Organisms used in the insertion studies included:

Bacteria:	*Escherichia coli* *Salmonella typhimurium* *Shigella sonnei* *Staphylococcus aureus* (coagulase positive)
Parasites:	*Endamoeba histolytica* *Ascaris lumbricoides* (viable ova) *Endolimax nana* *Necatur americanus*
Fungi:	*Histoplasma capsulatum* *Blastomyces dermatitidis* *Aspergillus fumigatus* *Geotrichum candidum*
Viruses:	Polio virus, Type II
Spirochetes:	*Leptospira philadelphia*

A total of 1,137 samples of bacteria, fungi, parasites, and viruses was inserted in 24 windrows in the second phase. In conformity with the first phase of the work, the samples were planted at various positions within the compost and withdrawn at intervals during the process.

Insertions of bacteria were made both in sealed ampules and in such manner that the cultures were in contact with compost. This was done to determine if bacteria could survive the process despite the elevated temperatures and the possible presence of antibiotic substances. Other organisms were inserted in test tubes with screw caps.

In the insertion studies, all bacterial samples in the form of impregnated discs in contact with compost removed after 25 days in the windrows were rendered nonviable. All samples of *Shigella sonnei* and *Staphylococcus aureus* removed at 14 days and thereafter were nonviable.

Samples of *S. sonnei* inserted in sealed ampules removed after 25 days were nonviable and those of *S. aureus,* similarly prepared, removed after 35 days were destroyed. Only two samples of 18 of *S. aureus* in ampules removed between 25 and 29 days in the compost were positive.

One sample of *Salmonella typhimurium* in an ampule was positive after 49 days and showed a reduction of cells from 2×10^8 to 1.6×10^2. This was a sample inserted and kept in the bottom 6 inches of a windrow where temperatures are lower. On turning, the compost was moved from the bottom but the sample was reinserted in this cooler part of the windrow. Of the 38 samples of the fungus *Histoplasma capsulatum,* inserted in capped test tubes, three were found to be positive on withdrawal.

Withdrawals were made from the seventh day to the 28th day. One positive had been in the compost for 14 days, another 24 days, and the third for 26 days. The 14th and 24th day samples had been kept at the midpoint of the composting mass and the 26th sample in the outer layer at the 2 inch depth. All samples of the fungus *Blastomyces dermatitidis,* withdrawn at intervals up to the 28th day, were rendered nonviable.

Thermophilic fungi such as *Aspergillus fumigatus* are commonly associated with the composting of various types of solid wastes. On-site observations showed that *A. fumigatus* could be consistently isolated from composting material at Johnson City at all stages of the process. In the insertion studies it survived for 28 days in capped test tubes. Inserted samples of *Geotrichum candidum* survived for 24 days.

All samples of *Endolimax nana* and *Endamoeba histolytica* were destroyed after 8 days' exposure. All samples of *Leptospira philadelphia,* withdrawn at from 2 to 9 days of exposure were deactivated. Ova of hookworm *(Necatur americanus* or *Ancylostoma duodenale)* disintegrated after 7 days' exposure in the compost.

As with the occurrence studies, the insertion studies show that some morphologically intact parasite forms *(Ascaris, Trichuris, Necatur, Ancylostoma,* and *Hymenolopis* species) persisted to the end of the composting process. Facilities were not available at ETSU and the Tennessee State Department of Health for a determination of the viability of these parasite forms.

Of the 66 samples of an enterovirus (Polio II) assayed for active virus particles after remaining for various lengths of time in the composting mass, one 3 day and one 14 day sample were found positive. The number of active virus particles recovered from each of the two positive samples amounted to 1% of the original concentration inserted. No virus sample was positive after exposure for more than 14 days. The windrow temperatures recorded for the two positive samples at the time they were withdrawn were 133°F (56°C) and 147°F (64°C).

Concurrently with the studies carried on by the Public Health Service and the research done under contract by Gaby of East Tennessee State University, Morgan conducted a study of the survival of *Mycobacterium tuberculosis* in windrows at the Johnson City Plant. The organism used was an avirulant *M. tuberculosis* var. *hominus,* obtained from Trudeau Institute, Inc.

The insertion technique was used and the samples were planted in the compost during fall, winter, spring, and summer months. Results revealed that *M. tuberculosis* was normally destroyed by the 14th day of composting where the average temperature was 149°F (65°C). In all cases the organisms were destroyed by the 21st day. In the study with one windrow, all *M. tuberculosis* organisms were killed by a temperature of 140°F (60°C) or less. The negative results obtained in the search for pathogenic bacteria (such as Salmonella, Shigella, and Staphylococcus) in compost would indicate that windrow conditions at Johnson City will destroy such pathogens.

The results obtained in the pathogen insertion studies, which were supported by the in-house work with *Mycobacterium phlei,* confirm this conclusion. Of the fungi, *Blastomyces dermatitidis* samples withdrawn up to the 28th day were rendered nonviable. Although one of the inserted *Histoplasma capsulatum* samples was found to survive 26 days of composting, the search for this fungus in the compost was not successful. *Aspergillus fumigatus,* occasionally pathogenic to humans, could be isolated in the compost itself through the process.

Histoplasma capsulatum has been isolated in garden soils and *Aspergillus fumigatus* is ubiquitous. The literature does not cite any references to fungal infection among sanitation workers in association with the disposal of solid wastes. Present data do not permit any estimate of the possible hazard of pathogenic fungi in this connection and the results at Johnson City do not warrant any restrictions on the use of compost due to these fungi.

Although morphologically intact parasite forms persisted through the process, the data shown by Belding show that the ova of *Ascaris lumbrocoides, Trichuris trichuria, Necator americanus, Ancylostoma duodenale,* and *Hymenolopsis diminuta* are killed at time-temperature conditions which are less severe than those which obtain in composting. The work of Scott in composting bears this out for Ascaris ova. It was concluded that the forms found were nonviable althought they were not tested with live animals.

Only two of 66 samples of enterovirus (Polio II) were found positive after 14 days in the compost. It is believed that these may have been recontaminated during assay as only very low counts were found and these organisms may be inactivated in 30 minutes at 122° to 130°F. No positive samples were found on withdrawal before 14 days.

Evidence from the ETSU studies, the in-house studies at Johnson City, and the work of others in the United States and Europe, indicates that properly processed compost of municipal waste and sewage sludge as used in gardening or agriculture does not constitute a public health hazard. For proper processing, it is imperative that all material in a given windrow be exposed to temperatures in the range of 122° to 130°F for a period of at least 7 days.

To accomplish this the material in the toe area must be picked up and piled on top or at the end of the windrow before turnings. At Johnson City the temperature of each windrow is taken weekly to be sure that the necessary temperatures are reached and held for 7 days. Actually, windrows which have not reached 140°F or over and remained for 7 days, are condemned at Johnson City in order to have a factor of safety.

Cellulolytic Activity in Composting

Cellulases were extracted by homogenizing 200 gram samples of compost in 2 liters of cold phosphate buffer (pH 7.0). The homogenates were clarified by centrifugation, and cellulase activity in the supernatant fluid was determined by measuring the liberation of reducing sugars (as cellobiose) from acetate-buffered solutions of carboxymethylcellulose. The pH, temperature, ionic strength, and substrate concentration optima were determined for this reaction. The total free cellulase content of compost was measured in relation to the age of the compost to determine the time of maximal production of cellulolytic enzymes. The cellulolytic flora which produce these cellulases in compost were isolated on microcrystalline cellulose-mineral salts agar and identification of these organisms was initiated.

Detectable cellulase activity (as measured by its ability to liberate reducing sugars from carboxymethylcellulose) increased tenfold at a logarithmic rate during the 49 day composting cycle. Concurrently, the cellulose content of the compost decreased from 50 to 30%. The pH and temperature optima for the cellulolytic reaction were consistently 6.0 and 149°F (65°C), respectively.

Variation of ionic strength between 0.1 and 1.0 had little effect on the velocity of the cellulase reaction. Maximal reaction velocity was achieved with a carboxymethylcellulose concentration of 2.5%, a concentration which is near the limit of solubility.

Three species of cellulolytic flora (designated as C-1, C-2, and C-3) were isolated from compost homogenates. C-1 resembled *Aspergillus fumigatus,* grew rapidly at 95° to 113°F (35° to 45°C), and caused intense clearing of the cellulose agar within 5 days. C-2, a Gram-variable spore-forming rod, grew rapidly at 95°F (35°C), slowly at 113°F (45°C), and produced diffuse zones of clearing on cellulose agar. C-3 appeared to be a thermophilic actinomycete (rapid growth at 125° to 131°F); it produced wide zones of cellulose clearing around colonies within 72 hours.

During the 49 day composting process, total cellulolytic activity increased tenfold while the cellulose content of the compost decreased from 50 to 30%. Consideration of temperatures which exist in the windrows would indicate that the greatest portion of cellulolytic acitivity is produced by the thermophilic actinomycete (C-3).

This organism, which grows rapidly at 131°F (and has an upper growth limit of 140°F) is the only species isolated which seems compatible with windrow temperatures. This actinomycete might be profitably exploited in various cellulose decomposition or transformation processes.

FLY POPULATION COUNTS

As mentioned in the section on operations, an extensive fly problem developed in the plant area in the summer of 1967. Flies in all stages of their life cycle came to the plant with the refuse being delivered in compaction trailers. Adult flies were widespread over the plant area and were particularly concentrated on windrows just being formed. Although many flies came in with the refuse, there was some breeding in the plant at the foot of windrows. Fly counts were taken in October 1968 to determine the comparative attractiveness of the storage area and several other areas. The storage area contained compost from 42 days old to more than 1 year old. The findings indicated that the stored compost had no attractiveness under the conditions which obtained. As would be expected, the greatest number was found where raw, unground refuse was being handled. The observations were made at the end of the fly season and may not be valid for a period of higher populations. At no time in 1968 or 1969 did the fly population reach a nuisance level at the plant.

CHEMICAL AND PHYSICAL CHARACTERISTICS

Moisture Content of Raw Refuse

With improved methods of sampling for moisture determinations, the dry weights of a number of windrows (11) were determined in the course of the work on the weight and volume losses. Knowing the weights of the same material as received from the city, the moisture content of raw refuse was determined. The moisture content by wet weight as received ranged from 22 to 61% over the period August 8 to December 6, 1968, with an average of 39%. This is in the neighborhood of the 35% which has been quoted on occasion.

Weight and Volume Losses in Composting

From 20 to 30% (by wet weight) of the incoming refuse is not compostable and is removed for burial in a landfill. Grinding then reduces the volume of the refuse retained for processing. As digestion proceeds, a weight loss in the form of the two principal products of decomposition, carbon dioxide and water, occurs. This amounts to between 20 and 30% of dry weight.

TABLE 7.6: WEIGHTS AND VOLUMES OF INCOMING COMPOST

	Refuse, as Received in Compaction Trailer	Noncompostables, Uncompacted	Picked, Ground Refuse, with Water Added	Compost in Windrow at End of Process
Wet weight, tons	100	28.8	111.7	50.4
Volume, cu yd	433	168.4	236.5	172.6
Moisture, percent by wet weight	39	39	60	30
Density, lb per cu yd at given moisture content	463	342	945	584
Density, expressed as the dry weight in lb per cu yd	282	209	378	409

A study of seven batches (windrows) of refuse through the 42 day composting process at Johnson City showed the relationship of incoming weights and volumes to those of the compost. In this study summarized in Table 7.6, the amount of noncompostables was somewhat higher than the average of 26% observed at this plant.

This study shows a compost yield of 50.4% by wet weight and 40% by volume. It is assumed that the compost will go down on the field at the desired 60% moisture and that the finished compost will be dried to 30% moisture.

Elemental Composition

Analyses for certain elements were performed on samples of compost 42 days or older. Table 7.7 shows the average values found, on a dry weight basis. Table 7.8 gives the results for individual analyses made for six important elements. The values are expressed as percentage of dry weight unless otherwise noted.

Nitrogen in fresh ground refuse ranged from 0.58 to 1.01% by dry weight, averaging 0.85%. Carbon ranged from 34.4 to 43.6% of dry weight, averaging 39.8.

For compost 42 days old or older, the nitrogen content was found to be between 0.85 and 1.07% of dry weight, averaging 0.93%. Table 7.9 shows the nitrogen content of some organic materials and soil for comparison with that of finished compost.

Analyses for Carbon/Nitrogen Ratios

A total of 37 windrows of different ages (0 to 130 days) was sampled and analyzed for carbon and nitrogen. Figure 7.8 shows the average carbon to nitrogen (C/N) ratios of composting refuse containing 3 to 5% sludge by dry weight for various ages.

Ranges are shown but the data, taken from 24 windrows, does not represent a follow through of all these windrows from 0 to 80 days. The initial C/N ratios of a group of seven windrows showed an average of 44.4. At 42 days of composting a similar group of five windrows showed an average C/N of 33.9 and at 49 days two of these showed an average of 31.9.

Figure 7.9 shows the decrease of the C/N ratios with age in a windrow especially sampled for this purpose. This windrow did not contain sludge solids. The initial C/N was 44.4 and at 49 days the ratio was 33.0. The similarity of Figures 7.8 and 7.9 will be noted.

The C/N ratios at the beginning and end of the composting on the field were determined for several windrows containing additives (Table 7.10).

TABLE 7.7: ELEMENTS IN FINISHED COMPOST

Element	Average (Percent Dry Weight)	Range (Percent Dry Weight)
Carbon	33.01	26.23 - 37.53
Nitrogen	0.93	0.85 - 1.07
Potassium	0.30	0.25 - 0.40
Sodium	0.42	0.36 - 0.51
Calcium	1.55	0.75 - 3.11
Phosphorous	0.26	0.20 - 0.34
Magnesium	1.61	0.83 - 2.52
Iron	1.18	0.55 - 1.68
Aluminum	0.94	0.32 - 2.67
Copper	less than 0.05	--
Manganese	less than 0.05	--
Nickel	less than 0.01	--
Zinc	less than 0.005	--
Boron	less than 0.0005	--
Mercury	not detected *	--
Lead	not detected *	--

* Lower limits of detection: Mercury, 0.005 percent; lead, 0.05 percent. Neither was found above these levels in any samples tested.

TABLE 7.8: CONCENTRATION OF CERTAIN TRACE ELEMENTS IN SCREENED COMPOST

Windrow and age	Iron	Boron	Sodium	Potassium	Phosphorous	Calcium
17D - 42 day*		<2ppm	0.50	0.30		0.75
17D - 49 day*	1.38		0.40	0.29	0.22	1.12
5E - 42 day†	1.19		0.42	0.28		0.92
2E - 42 day‡	0.89	<2ppm	0.36	0.25	0.20	2.46
5E - 42 day§	1.07	<1ppm	0.43	0.25		1.89
Stockpile 1						
- 56 day¶	0.97		0.51	0.33	0.29	1.85
22-23E - 77 day‡	1.24	<2ppm	0.43	0.29	0.25	1.40
22-23E - 130 day‡					0.34	
Stockpile 2						
8G - 36 day¶	1.35	<1ppm			0.27	2.39
with NH_4NO_3 added					0.23	
5E - 154 day†	1.68	<1ppm	0.39	0.29	0.29	3.11
24-25E - 130 day§	1.47	<1ppm	0.43	0.32	0.24	2.29
8G - 42 day§	0.55		0.36	0.40	0.28	1.13

* Contained sewage sludge in an undetermined amount
† Contained 12 percent sewage sludge solids.
‡ Contained 2 to 5 percent sewage sludge solids.
§ This portion of 5E contained 2 to 5 percent sewage sludge solids.
¶ Contained sewage sludge solids.

TABLE 7.9: NITROGEN CONTENT OF SOME ORGANIC MATERIALS AND SOIL

		% Nitrogen (dry weight basis)
(1)	Finished Compost	0.93
(2)	Sewage Sludge (raw)	2.58
(3)	Chicken manure	2.61, 2.40
(4)	Cow manure	3.42
(5)	Peat moss	1.91
(6)	Leaves (hardwood)	1.34
(7)	Pepper canning wastes	2.82
(8)	Tobacco stalks	1.41
(9)	Soil	0.1 - 0.3 (approximately)

FIGURE 7.8: AVERAGE CARBON TO NITROGEN RATIO OF 24 WINDROWS CONTAINING SEWAGE SLUDGE

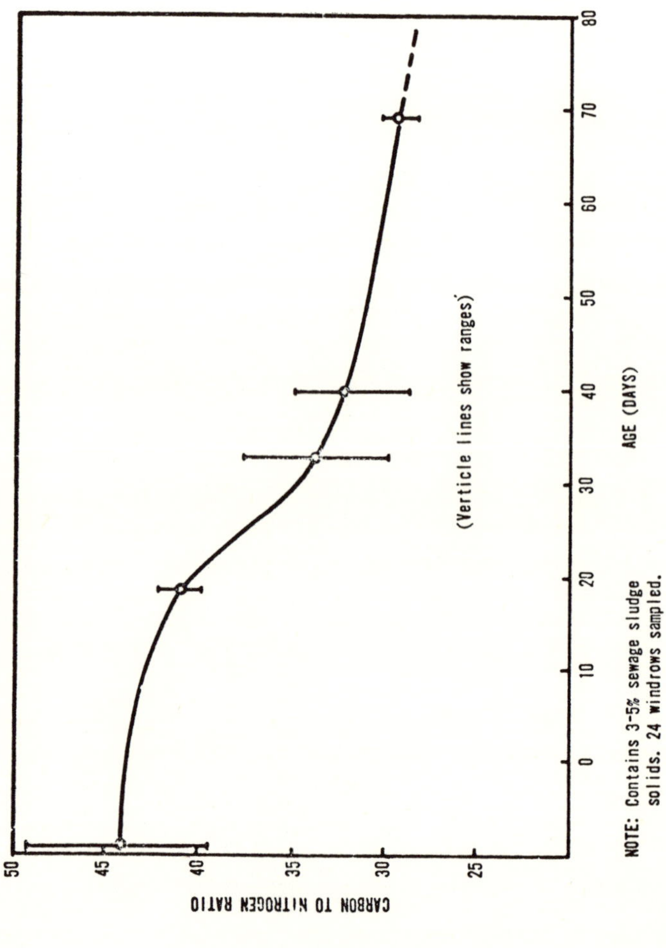

Source: PB 214,056; 1972

FIGURE 7.9: CARBON TO NITROGEN RATIO OF A WINDROW WITHOUT SEWAGE SLUDGE

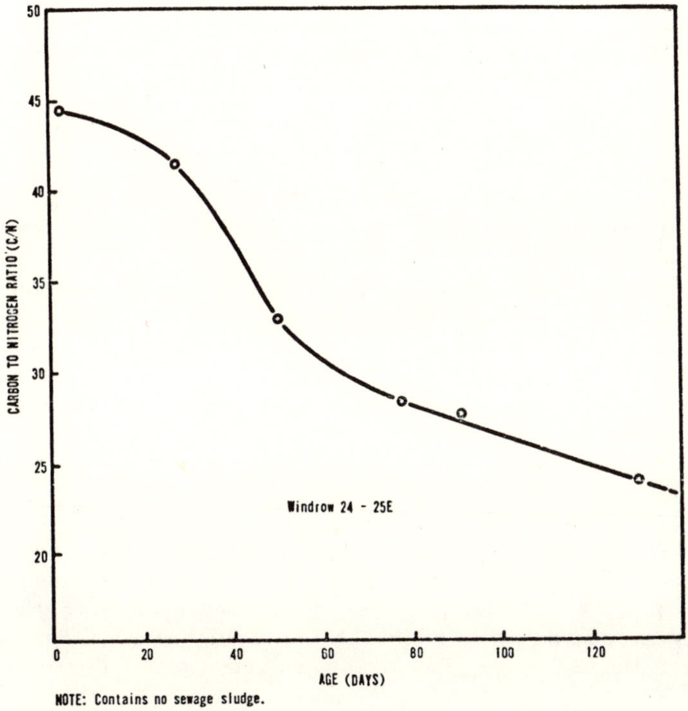

Source: PB 214,056; 1972

TABLE 7.10: CARBON/NITROGEN RATIOS OF COMPOST CONTAINING ADDITIVES

Windrow and Age (days)		Additive	C/N Ratio
13G	1	Paunch manure	31.2
	42		28.5
19G	1	Chicken manure	25.8
	42		24.0
8G	1	Urea-ammonium nitrate	9.8
	42		12.9
17G	1	Urea-ammonium nitrate	17.7
	42		22.4

(continued)

TABLE 7.10: (continued)

Windrow and Age (days)		Additive	C/N Ratio
1H	43	Urea-ammonium nitrate	29.6
	56		22.8
30G	54	Animal blood	22.5
33G	51	Pepper canning waste	26.5

Windrows to which were added paunch manure and poultry manure showed lower initial C/N ratios. The decreases in the ratios while composting, however, were not as great as in normal windrows. The 42 day ratios were 28.8 for the material containing paunch manure and 24 for the chicken manure-refuse mixture.

The C/N ratios for two windrows to which urea-ammonium nitrate was added showed an increase rather than a decrease after composting. The rise was due to the loss of nitrogen as ammonia as a result of mass action from (a) the increase of pH in the system caused by the addition of urea-NH_4NO_3, and (b) shifting of the equilibrium $NH_4^+ + OH^- \leftrightharpoons H_2O + NH_3\uparrow$ to the right due to an increase of NH_4^+ and OH^- ions.

Analyses for Chemical Oxygen Demand

A total of 22 samples of compost of different ages (0 to 168 days) was analyzed for chemical oxygen demand (COD). The results showed a steady reduction with age from about 900 mg/g for fresh refuse to about 750 mg/g at 56 days of age. A sample taken from compost 168 days old showed a COD of about 300 mg/g. The COD can be used as a gauge of the degree of decomposition of refuse.

Cellulose, Starch and Sugar Content

The average cellulose content of refuse at Johnson City was found to be 49% on a dry weight basis. After composting for 28 days, it was 43% and at 49 days the content was 31%. The average initial starch content was 4.0% and dropped to less than 1.0% after 28 days of composting. The average sugar content at the start of composting was 0.8% and dropped to less than 0.1% within 14 days.

Calorific Value of Refuse and Compost

At Johnson City, the calorific values for composite samples of composting refuse containing 3 to 5% sludge solids were found to be initially 4,077 calories/gram, diminishing to 3,669 on the 49th day of composting and to 2,947 calories/gram for compost 1 year old.

COST STUDIES

A system of cost accounting for plant operations and maintenance was developed to provide accurate cost data for each phase of plant operation for appraisal of windrow composting as a method of solid waste management. Plant operations and activities were divided into various categories or units with an account number each. Cost data used to report the project costs and provide the basis for projections to other windrow plants were obtained from these monthly, quarterly, and year-end financial statements of operations.

Capital Cost: The total construction cost for the plant, including the modifications made since startup, is $958,375. On the basis of per ton daily capacity the actual capital investment costs were $18,580. Capital cost calculations assumed straight line depreciation over 20 years of buildings and equipment (excluding land) and bank financing at 7½% over 20 years. On a per ton refuse processed basis, the capital investment cost is $12.98 (at 34 tons per day in 1968).

The average rate at which the refuse has been received at the plant would permit the processing of 52 tons in a normal processing day (6½ hours for processing and 1½ hours for cleaning operations). Projecting the actual costs to the 52 tons per day basis, the capital investment cost would be $6.88 per ton refuse processed.

Operating Cost: Actual costs for operating the composting plant in 1968 were $18.45 per ton refuse processed. The nature of the operations (research and development) and the inability of the Johnson City municipality to deliver enough refuse for operation at full capacity are some reasons for this high unit cost. A cost of $13.40 per ton refuse processed was projected for operating this plant at 52 tons per day. Some modifications were made for the research work being conducted.

The construction cost of $958,375 for the plant is subject to some qualifications. A high proportion (38% of plant cost) is in buildings, partly because of the multistory design, with equipment installed on the second and third floor levels.

More ground level floor space and simpler framing, as used in common mill buildings, with installation of equipment independently of the structure, would have permitted less expensive construction. A case in point is the 150 ton per day plant at Gainesville, Florida, where the cost of the buildings, estimated at $150,000, is approximately 11% of the total plant investment.

Labor Cost: With respect to the plant operating costs in 1968, labor constituted about 75% of this cost. Based on 50 tons of refuse processed per day, 1969 labor costs amounted to 78% of operating expenses.

Cost Data Projected to Other Plants: Estimates of the capital, investment, and operating costs for various capacity windrow composting plants were developed based on the actual costs recorded during the report period at the USPHS-TVA plant. The projections indicate that the total yearly costs for various size windrow composting plants will range from $19.77 per ton refuse processed for a 50 ton per day plant to $10.71 per ton refuse processed for a 200 ton per day plant on two shifts. The cost estimates include equipment for processing sewage sludge. Also included are costs for land, depreciation, and debt service. Because use of compost is normally seasonal, the estimates include land area for storage of 6 month's production of compost in rectangular piles 15 ft high.

Land costs were assumed at $800 per acre, consistent with land values near Johnson City. If land costs were assumed at $1,500 per acre, the capital cost for the 100 and 200 ton plants on two shifts would increase by no more than one cent per ton of refuse processed. For the 100 ton plant on one shift, the increase would be about three cents per ton of refuse processed.

Plant Income: None of the compost produced has been sold. Also, salvaging of potentially salable materials has not been practiced. Therefore, the project has not obtained cost data with respect to potential plant income from such sources.

DEMONSTRATION AND UTILIZATION

None of the compost produced at the Johnson City plant has been sold. Prior to March 1969, the Bureau of Solid Waste Management had asked TVA to restrict the uses to which it was put pending the outcome of pathogen survival studies. These restrictions and the lack of a finished product has limited the activity of the Division of Agricultural Development, TVA, in its utilization studies.

The TVA agriculturist assigned to the project has, however, been active in setting up demonstrations where the unfinished material could be used and where owners of the sites agreed to abide by the restrictions. The demonstration areas are on public lands or the land of private owners who have agreed to allow the agriculturist to supervise the application and to follow the progress of the plantings.

Due to the troubles experienced with the final grinding and screening equipment which was not completed until February 1969, the bulk of the material used in Fiscal Year 1969 was unfinished (unreground or screened). It represents 80% of the compost produced in the year.

The demonstration sites include tobacco, corn, gardens, grass or sod establishment, erosion control and reclamation, orchards, shrubs and flowers, golf courses, soybeans, and miscellaneous.

TVA experimental test plots are in corn and grain sorghum, each involving 52 test plots, 12 x 30 ft each, in which various application rates are being examined, some with fertilizers and some without. Another concerns an evaluation of compost on Bermuda grass. Preliminary results of the experiments are described below. The rates of application on the demonstration plantings range from 10 to 100 tons per acre for corn and 5 to 30 tons per acre for tobacco. The agriculturist hopes to evaluate the merits of application rates of 4 to 200 tons per acre and various rates of fertilization in several seasons over the next 3½ years. Three other demonstrations are described in Chapter 3 page .

Corn Grain Research Project

This project involves the use of refuse-sludge compost in the growth of corn on agricultural soils. Application rates in fall or spring ranged from 0 to 200 tons per acre. The purpose of such a range in application rates was to determine the soil and crop improvement resulting from various amounts of compost and the maximum amount of compost that can be applied before the effect is adverse or deleterious.

The fall applications of 4, 8, 50, 100, and 200 tons per acre of unscreened compost were made in November 1968, and plowed under shortly thereafter. Spring applications were made in April 1969, and disked into the soil. Nitrogen rates of 80 and 160 lb per acre were also compared alone and with 8 tons of spring applied compost. There was a total of 52 test plots with each plot receiving phosphorus and potassium at rates adequate for maximum plant growth. The corn was planted on May 2, 1969.

There was a slight reduction in germination on the plots that received the three highest rates of compost (50, 100, and 200 tons per acre) with the lowest percentage germination coming on the 200 ton per acre rate. One factor that contributed to this reduction in germination was the inability because of the bulkiness of the compost to prepare a firm seedbed, especially on the plot receiving the 200 ton per acre rate.

Other factors that affected the overall number of plants on all plots were (a) the crop was damaged somewhat by birds, (b) extremely cool temperatures occurred during the germination period, and (c) there was also some damage caused by hail.

Nitrogen deficiencies developed in all plots that did not receive supplemental inorganic nitrogen. Early growth was extremely poor on the plots that received the three heaviest compost applications; however, toward the middle of the growing season, the corn on these plots began to make some progress.

The first noticeable growth response came on the 50 ton per acre treatment followed in sequence by the 100 and 200 ton per acre rates. The final results indicated that these three treatments gave favorable results when compared to the plots receiving the 80 and the 160 lb per acre nitrogen treatments with no compost. It was also noted that the corn on the 200 ton per acre plot remained green longer than any other treatment.

These early results would seem to indicate that the heavy application of compost had a definite effect on the amount of nitrogen being released to the soil in a form readily available to the plant, the heavier the application of compost, the longer it took for the corn to respond. It was approximately 7 months from the time the compost was plowed under in the 200 ton per acre treatment before there was any visual growth response that could be considered a normal growth pattern. Final yield results of this first year observation showed that grain yields increased from 55 bushels per acre without nitrogen or compost to 76 bushels with 80 lb of nitrogen or with 100 tons of compost and to 90 bushels per acre with 160 lb of nitrogen plus 8 tons of compost. Four to eight tons per acre of compost alone resulted in slight, if any, increase in corn yields.

The value of compost on corn in terms of increased yields ranged from a -\$1.75 per ton on spring applied compost at a rate of 4 tons per acre with no supplemental nitrogen to a +\$3.18 per ton on spring applied compost at a rate of 8 tons per acre with 160 lb of supplemental nitrogen per acre.

The value of compost when applied at a 200 ton per acre rate with no supplemental nitrogen was \$0.08 per ton, however, this does show that large amounts of compost can be utilized on agricultural soils with positive results. Additional studies will determine the residual effect of high rates of application of compost over a period of years.

It was apparent during the first year of testing that a combination of compost plus inorganic fertilizer produced greater yield than either compost or fertilizer alone. The response from the corn on the plot that received compost plus 160 pounds of nitrogen was evident throughout the growing season as the corn stalks appeared greener and stronger than stalks from other plots.

When inorganic fertilizer was used without compost, the yield on the 80 lb per acre of nitrogen and the 160 lb per acre of nitrogen treatments was approximately the same which would indicate that the excess over 80 lb per acre of nitrogen was not efficiently utilized by plant uptake.

This situation can be partly attributed to the relatively small number of plants per plot; however, when 8 tons per acre of compost was used in conjunction with 160 lb per acre nitrogen, there was an approximate 20% increase in yield over the 160 lb per acre of nitrogen alone or the 8 tons per acre compost with 80 lb per acre of nitrogen.

It would appear that the compost had somewhat of a synergistic effect on the inorganic nitrogen since it is very doubtful that 8 tons of compost per acre would supply enough nitrogen or moisture to bring about such an increase in yield.

There are some preliminary conclusions drawn from one year of testing. More valid information can be gathered from results obtained from four or five years of continuous testing. Table 7.11 presents a summary of yield data obtained from corn research at Johnson City during 1969.

TABLE 7.11: EAR CORN YIELDS, ACRE BASIS

Treatment	Stalks	Ear Corn Pounds	Ear Corn Bushels
1 - None	8230	3570	55
2 - 80 N	8470	4960	76
3 - 4T-Spring	8230	3235	49
4 - 160 N	8230	4750	73
5 - 8T-Spring	8710	4325	66
6 - 8T-Fall	8230	3810	58
7 - 4T-Fall	7745	3900	60
8 - 8T-Spring + 80 N	8710	4900	75
9 - 8T-Spring + 160 N	8470	5870	90
10 - 8T-Fall	8230	3780	58
11 - 50T-Fall	7745	4535	69
12 - 100T-Fall	7745	4990	76
13 - 200T-Fall	6775	4295	66
L.S.D., 5% Level	N.S.	855	13

Compost contained 40% moisture; application rates are on a dry basis. N was applied in spring as ammonium nitrate. Pounds of ear corn are as harvested (Av. of 14.5% moisture); bushels are on the basis of 15.5% moisture (70 pounds per bushel). The corn variety was Funk's G-5757.

Use of Compost for Sorghum Production

The source of the following information is Experiment 56, National Fertilizer Development Center, Muscle Shoals, Alabama. Yields of dry sorghum forage increased from 5 tons per acre without compost or N to 7.5 tons with 82 tons of compost and to 8 tons with 160 lb of applied N. Thus, the value of the compost in terms of the N it supplied was very low.

Less wilting during dry periods on plots receiving large applications of compost indicated that it increased the moisture holding capacity of the soil. The results also indicate that agricultural land will accept large amounts of compost with small positive yield effects.

Procedure: Unground compost from Johnson City, Tennessee, was applied in fall, spring, or combination applications for Funk's 101F forage sorghum at total rates ranging from 6 to 82 tons per acre. Fall applied compost was plowed under shortly after application, while spring applications were incorporated by disking. Compost rates were compared with 80 to 160 lb per acre, with three treatments supplying both N and compost and a check treatment with no N or compost. Chemical components in the compost were as follows (percent, dry weight basis).

Application Time	N	P	K	C	Ca	Na	Mg	S	Zn
Fall 1968	1.2	0.24	0.80	34.2	3.6	0.49	0.49	0.4	0.13
Spring 1969	1.3	0.40	0.96	26.8	6.4	0.82	0.87	0.4	0.15

P and K were applied at rates thought to be adequate for maximum plant growth. Forage sorghum was planted on April 23, and harvested on July 28 and October 15.

Results and Discussion: Forage yields, percentages, N content and N uptake are given in Table 7.12. Total yields of dry sorghum forage increased in a generally linear fashion with higher compost rates. Yields ranged from 10,814 lb per acre on the check plot to 15,032 with 82 tons of compost per acre.

In contrast, 80 lb of N produced 13,850 lb of forage, while 160 lb produced 16,384 lb per acre. The highest yield, 18,823 lb per acre was produced by 160 pounds of N and 40 tons of compost. Forty tons of compost produced as much forage as 80 lb of N, but no compost rate was as good as 160 lb of N. In all cases, compost plus N produced more forage than either material alone. At the first harvest, N contents of the forage were much higher with N alone than with compost alone, but differences were small at the second harvest. N uptake for similar yields was less from compost treated than from N fertilized plots.

TABLE 7.12: RESPONSE OF FORAGE SORGHUM TO APPLICATIONS OF COMPOST AND N

Compost Rate		N	Dry Forage, Pounds/A			N Content, %		N Uptake, Pounds/A			
Fall	Spring	Rate	Cut 1	Cut 2	Total	Cut 1	Cut 2	Cut 1	Cut 2	Total	Increase over no N or compost
Tons/A		Pounds/A									
0	0	0	8155f	2659e	10814g	.63	1.08	51.3	29.2	80.4g	-
0	0	80	10205c	3643de	13850de	1.08	.91	109.9	33.7	143.6d	63.2
0	6.2	0	8696ef	2943e	11640fg	.66	.95	57.3	27.9	85.1g	4.7
0	0	160	11306f	5076b	16384b	1.33	1.13	150.6	58.1	208.8ab	128.4
0	12.4	0	9214de	3434e	12649e	.71	1.03	65.5	36.0	101.5f	21.1
8	0	0	8736ef	3158e	11895fg	.65	1.00	55.9	31.5	87.3fg	6.9
4	6.2	0	9035def	3411e	12447f	.68	.90	61.6	30.7	92.3fg	11.9
0	12.4	80	11717b	4138cd	15856bc	1.10	.99	128.0	41.0	169.0c	88.6
0	12.4	160	12158ab	5729ab	17888a	1.15	1.06	139.1	61.0	200.1b	119.7
8	12.4	0	8871def	3531d	12403f	.72	1.06	63.7	37.3	101.0f	20.5
16	24.8	0	9483cde	4340c	13825de	.78	.97	73.2	42.0	115.3e	34.9
32	49.6	0	9773cd	5258b	15032cd	.88	1.11	86.4	55.1	141.4d	61.0
16	24.8	160	12670a	6151a	18823a	1.20	1.11	150.0	67.4	217.4a	137.0

Note: In all yield tables means with the same letter are not significantly different at the 5 percent level.

The N content and uptake data indicated that this was probably due largely to luxury uptake of N on fertilized plots. Observation of the sorghum during dry periods indicated that compost contributed some moisture holding capacity to the soil. There was always less leaf curling on plots treated with the higher compost rates.

A slight stand reduction occurred on plots receiving 50 tons in the spring. It was not possible to incorporate this amount of compost well enough to provide the firm seedbed needed for good germination.

Value of compost per ton in terms of its contribution to yield increase was lower with increasing application rates. Dry matter yield increases per ton of compost ranged from 255 lb at the 6 ton rate to 52 lb at the 80 ton rate. The yield contributions were similar with and without N additions.

Assuming that sorghum forage is worth $10/ton on a green weight basis, compost was worth about $2.20/ton at 8 to 12 tons/acre application rates and approximately $1.10 at rates of 20 tons/acre or more. If, on the other hand, the value is based on the amount of N needed to give equal yield, it was worth about $0.13/ton.

As with the Bermuda grass experiment (No. 57), this experiment showed that large amounts of compost can be disposed of by application on agricultural land without apparent yield reductions. The compost used in these two experiments was unground and contained plastic, both film and dense type, fairly large pieces of glass and metallic waste. It would have been completely unsuitable for application to grazing or hay land and was objectionable from an aesthetic viewpoint for most surface application uses.

Evaluation of Compost on Common Bermuda Grass

The source of the following information is Experiment 57, National Fertilizer Development Center, Muscle Shoals, Alabama. Yields of Bermuda grass forage increased from 2.9 tons with no compost or N to 3.5 tons with 12 tons per acre of compost. The corresponding yields with 160 lb per acre of N applied were 5.0 tons without compost and 5.2 tons with 12 tons per acre of compost. The increases for compost were of insufficient value to pay for application costs.

Procedure: Compost was topdressed on common Bermuda grass sod in November 1968, at rates of 0, 4, 8, and 12 tons per acre. In 1969, N rates of 0 and 160 lb per acre were superimposed on plots of each compost rate. Half the N was applied in April and half after the second harvest. The grass was cut four times. Results are presented in Table 7.13 on the following page.

Results and Discussion: Without added N there was a slight but consistent increase in total yield with increasing compost rates ranging from 5,772 lb per acre with no compost to 7,045 lb with 12 tons per acre of compost. Where N was applied the 4 ton per acre compost application resulted in a slight reduction in yield, and 8 and 12 tons per acre in slightly increased yield. The increase was greater with 8 than with 12 tons per acre of compost.

Although there were inconsistencies in yield response to compost, the 8 ton per acre rate at both N levels resulted in approximately 900 lb additional forage per acre. At $25.00 per ton as the market value for Bermuda grass hay, this compost would have been worth $1.25 per ton spread on the field.

This would hardly pay spreading costs let alone production and transportation expense. However, this data does indicate that it might be possible to dispose of significant amounts of compost by spreading it on grasslands providing that it did not contain solid material harmful to livestock.

TABLE 7.13: RESPONSE OF COMMON BERMUDA GRASS TO COMPOST AND NITROGEN

Treatments		Forage harvested, pounds				
Compost, T./A	N, Lbs/A	Cut 1	Cut 2	Cut 3	Cut 4	Total
4	160	2899	2177	1731	3290	10,098a
4	0	1663	1826	504	2644	6,639bc
8	160	3017	2533	2031	3590	11,172a
8	0	2025	1943	574	2211	6,755bc
12	160	2529	2696	1845	3444	10,515a
12	0	1628	1992	721	2702	7,045b
0	160	3156	2048	1961	3146	10,312a
0	0	1732	1436	535	2069	5,772c

Note: In all yield tables means with the same letter are not significantly different at the 5 percent level.

Illinois Composting Applications

The material in this chapter is from PB 223,752, October, 1972 written by D.L. Eberhardt and W.O. Pipes.

BIOLOGICAL BASIS OF COMPOSTING

Composting is a biological process depending upon living organisms. The life is in the form of saprophytic microorganisms, largely bacteria and fungi which can be found in densities of billions of organisms per gram of composting material. Dead and living microorganisms can make up over 25% of the weight of compost. Table 8.1 lists some of the types of bacteria and fungi which have been found to grow in compost. Actinomycetes are a type of bacteria which grow in a filamentous form like fungi.

TABLE 8.1: REPRESENTATIVE GENERA OF MICROORGANISMS FOUND IN COMPOST

Fungi		Actinomycetes	Bacteria	
Absidia	Mucor	Micromonospora	Aerobacter	Micrococcus
Aspergillus	Nigrospora	Nocardia	Bacillus	Myxococcus
Botrytis	Oospora	Streptomyces	Cellumonas	Proteus
Byssochlamys	Penicillum	Thermoactinomyces	Chondrococcus	Pseudomonas
Cephalosporium	Rhizopus		Clostridium	Sarcina
Chaetomium	Sporotrichum		Diplococcus	Streptococous
Coprinus	Stilbella		Escherichia	Thiobacillus
Fusarium	Thermoascus			
Geotrichum	Torula			
Humicola	Trichoderma			

The microorganisms utilize the organic matter as a source of nutrition and energy. Microbial utilization of organic matter results in the production of carbon dioxide, water, new microbial cells, and many other compounds which are produced as by-products of the activities of the microorganisms. The physical and chemical changes which are caused by microbial decomposition include changes in the appearance, odor, and chemical composition of organic material.

Many different types of organisms are required to decompose the various types of organic materials found in solid wastes. This is true because no microorganism can decompose all types of organic matter. Different types of organisms utilize different types of organic compounds. Many microorganisms produce by-products which can be utilized by other microorganisms thus creating chains of utilization by which complex organic compounds can be broken down.

Decomposition occurs best when there is an adequate supply of nutrients. Carbon, nitrogen, oxygen, phosphorus, potassium, hydrogen, sulfur, iron, and magnesium plus 10 to 20 trace element nutrients are essential for microbial growth. For proper decomposition the amounts of the various nutrients should be balanced with the need for each. The result of a deficiency of one or more nutrients is a decrease in the rate and the extent of decomposition. Table 8.2 lists the amounts and sources of the major nutrients required by the microorganisms active in composting. Most organic wastes have adequate amounts of trace nutrients but many have a disproportionately large supply of carbon (e.g., municipal refuse) or nitrogen (e.g., packing house waste).

TABLE 8.2: NUTRITIONAL REQUIREMENTS AND SOURCES FOR MICROORGANISMS ACTIVE IN COMPOSTING

Element	Required Range (Percentage of Dry Weight)	Sources
Carbon (C)	40 - 80	Organic matter
Hydrogen (H)	1 - 10	Organic matter, water
Oxygen (O)	8 - 30	Organic matter, water
Nitrogen (N)	0.1 - 4.0	Organic matter, ammonia, nitrates
Phosphorus (P)	0.01 - 1.0	Organic matter, phosphates
Potassium (K)	0.001 - 0.1	Salts
Sulfur (S)	0.001 - 0.1	Organic matter, sulfates
Magnesium (Mg)	0.001 - 0.1	Organic matter, salts
Iron (Fe)	0.001 - 0.1	Organic matter, salts
Trace nutrients (Cu, Co, Mn, Mo, etc.)	less than 0.001	Salts

One of the best indices of the compostability of a material is the ratio of carbon to nitrogen. If the C/N ratio is much greater than 40, the rate of decomposition may be greatly reduced. In materials with high C/N ratios the carbon content has to be reduced before decomposition is completed. This reduction may be achieved through a slow initial decomposition process in which successive generations of organisms which utilize carbon as a source of energy oxidize the carbon and release it as CO_2. The nitrogen in the organisms which die is reused by the later generations which oxidize more carbon. This process of carbon reduction can require many generations of microorganisms. If compost with a C/N ratio greater than 20 is applied to soil, it can temporarily tie up the nitrogen in the soil by utilizing the nitrogen in the soil to complete the decomposition of the carbon compounds. However, composts with a large amount of carbon which is not readily available to microorganisms and therefore decomposes very slowly, such as composts from leaves or municipal refuse, can have final C/N ratios greater than 20 and still not rob nitrogen from the soil.

The environmental conditions in the decomposing organic material will affect the rate and type of microbial growth. Some important environmental factors are: availability of gaseous oxygen, concentration of water, pH, particle size, presence of toxic substances, and temperature.

If an adequate supply of gaseous oxygen is available, aerobic microorganisms will predominate and aerobic decomposition will occur. Conversely, if little oxygen is available, anaerobic microorganisms will predominate and anaerobic decomposition will occur. Anaerobic decomposition is slower than aerobic decomposition, results in only a partial breakdown of organic matter, and often produces disagreeable odors. Supplying air to keep the oxygen supply adequate during decomposition is a major consideration in most composting techniques due to the advantages gained by maintaining aerobic conditions. This is usually accomplished by turning a compost pile or windrow but in some processes it is accomplished by a forced draft. Oxygen has very limited solubility in water; thus, the required rate of oxygen supply is related to the moisture content of the composting material.

Water concentrations of 45% to 65% have been found to give the best results in composting municipal refuse. Most organic wastes will compost best with moisture contents within this range. Water concentrations less than 45% can result in an inadequate supply of water for the microorganisms while concentrations greater than 65% can result in an inadequate supply of oxygen yielding anaerobic conditions.

Generally, materials with high C/N ratios can contain more water without becoming anaerobic than materials with C/N ratios between 20 and 40 due to the slower rate of decomposition and oxygen utilization in most materials with high C/N ratios. The water concentration that best suits a particular process and organic waste material will vary and will have to be determined by trial and error experiments.

The factor, pH, is a measure of the acidity or alkalinity of a mixture; a pH of 7 being neutral, a pH of less than 7 being acid, and a pH of more than 7 being alkaline. Decomposition will occur most readily with a neutral pH because most microorganisms grow best under neutral conditions. Under anaerobic conditions significant reductions in pH due to the microbial production of acids can be expected. Under aerobic conditions a material that is initially neutral will usually experience a decrease in pH during the start of composting. This will normally be followed by an increase in pH which will result in a final state that is slightly alkaline. Most materials decomposing aerobically will stay within a pH range that is conducive to microbial growth thus eliminating any need for pH control. The addition of lime for pH control, a common practice in composting, is usually not necessary and can result in unnecessary nitrogen loss.

Particle size has a major effect on the rate of decomposition. The smaller the particle size the larger the surface area available to the microorganisms. Materials such as municipal wastes decompose much more readily if they are shredded initially to reduce particle size. The presence of substances that are toxic to microorganisms will generally decrease the rate of decomposition. However, waste materials that contain toxic substances will normally be compostable if the concentration of toxic materials is relatively small because some microorganisms are resistant to toxic materials.

The oxidation of carbon to CO_2 provides energy for microbial growth and some of this energy is converted into heat. When decomposition is rapid and concentrated, as it is in many composting methods, a large amount of heat is produced and a significant temperature rise can be expected. When the temperature rises to the 105° to 115°F range, most of the ordinary microorganisms can no longer grow. At this point, thermophilic bacteria and fungi take over. Thermophilic bacteria and fungi are microorganisms that survive best in temperatures between 105° and 170°F.

These microorganisms can cause a further increase in temperature to 160° to 170°F. The increase of temperature has two important effects: it produces a faster rate of decomposition, and it kills pathogens, weed seeds, and fly larvae. Thermophilic microorganisms do not grow well at temperatures above 170°F; therefore, temperatures in the compost will usually not go beyond this point. Most composting methods which are designed for rapid decomposition provide conditions which will support thermophilic decomposition.

The composting process is complete when the organic material will no longer support rapid microbial growth. In the case of thermophilic composting, the completion of the process is indicated by a reduction of temperature below the thermophilic range.

LEAF COMPOSTING

Since open burning has been restricted in Illinois, many communities are faced with a serious leaf disposal problem. The alternatives for leaf disposal are burning in closed incinerators, land-filling, and composting. Incineration has the obvious disadvantage of high cost and air pollution. Most municipal refuse incinerators cannot handle the additional load of leaves, particularly when leaves are too moist. Landfilling leaves also has disadvantages. Leaves do not compress well and thus consume large amounts of landfill space. Furthermore, leaves do not decompose rapidly under anaerobic conditions in the landfill and will cause long term settling problems over a period of years.

Composting is an alternative which is environmentally beneficial and in many cases could be economically practical. Under proper conditions leaves can be decomposed into a compost, called leaf mold, which serves well as a soil conditioner. Even when landfill is used for ultimate disposal of leaves, composting can eliminate much of the disposal problem because leaf mold is only 5 to 10% of the volume of the original leaves.

The degree to which leaf composting is a reasonable disposal alternative will be different for each community. The major factors that determine the practicality of leaf composting for a particular community are: (1) the size of the problem; (2) the plausibility of alternative disposal methods; (3) the availability of land for a compost site; (4) the community need for leaf mold; (5) the plausibility of leaf pickup from the community or leaf drop-off at the compost site; and (6) the environmental awareness of the community.

Communities with many trees will have large accumulations of leaves making composting more attractive than alternative methods of disposal. The cost of burning leaves in closed incinerators or of landfill disposal varies greatly from one community to another and obviously must be considered before composting is adopted. Leaf composting will be more attractive when the costs of the alternative methods of disposal are high. The spatial requirements for leaf composting are not excessive because the leaf windrows can be eight feet or more high. Most communities will have a suitable site. Communities which buy peat moss or topsoil for use on public parks and gardens can use leaf mold as a replacement for these materials and thereby offset part of the expense of composting.

Collection and transportation of the leaves to the composting site will be a major part of the cost of composting. Many communities already provide some form of leaf collection in the fall either by municipal or private collectors. In some communities with a high degree of environmental awareness, citizen cooperation and participation can be most helpful particularly in holding down the cost of the initial collection and transportation of the leaves to the composting site.

Leaf Collection

Many municipalities chose to provide leaf collection when leaf burning was banned. Leaves are usually collected either in bags or in piles at the street curbing. Bagging permits leaves to be collected with conventional refuse collection techniques but requires a major effort on the part of the community's residents, especially in heavily foliaged areas. Bagging also makes composting more difficult because leaves must be removed from bags before they can be composted properly. Loose leaf collection from street curbings has proven to be an effective means of leaf collection and is practiced by many municipalities, particularly by those with large quantities of leaves. With this method leaves that are raked into piles at the street curbing are picked up by any of a number of methods. Large vacuum units have worked well for curb pickup when leaves are not too moist. Front-end loaders are used by many municipalities when the leaves are too wet for pickup with a vacuum unit.

A jeep or tractor with a broom mounted in front has been found to be useful to push leaves into large piles which are more easily collected by either vacuum units or front-end loaders. Street cleaners can also be used effectively when small amounts of leaves are involved either early or late in the season or between vacuum or front-end loader collection during periods of heavy leaf accumulation. The best collection technique will be different for every community but much can be learned from those municipalities that have developed leaf collection systems through trial and error experimentation.

Leaf collection is the most costly step in the process of leaf composting. The delivery of leaves to the compost site by individual residents of a community is an alternative to municipal leaf collection. This is a possibility in municipalities where no leaf collection is provided or where a fee is charged for the collection of bagged leaves. A compost site available to the community for disposal of leaves can be an inexpensive alternative to providing community-wide leaf collection, but unless there is good citizen cooperation this method could result in many difficulties, such as an accumulation of refuse as well as leaves at the compost site.

Site Location

The compost site should be centrally located to minimize transportation costs. A site isolated from residences is preferable but a well maintained site will not be offensive and could be located in close proximity to residences. The site should have a sufficiently hard surface, either natural or paved, to permit vehicular access during wet weather. Drainage should be adequate to prevent puddle formation which can result in anaerobic conditions in the compost piles. Soil that allows good vertical drainage and/or the presence of a slight grade for horizontal drainage will prevent puddle formation.

When curb collection is provided, it has been estimated that one acre of composting area is required for each square mile of community if the home density is 800 to 1,000 per square mile and one half acre for each square mile of community if the home density is 400 to 600 homes per square mile. It is assumed that these estimates are for communities with average amounts of foliage.

When the compost site is to be used by the community it is important that it be accessible. Some communities have turned their leaf composting site into environmental centers where citizens can drop off recycleable materials such as paper, glass, and cans or can pick up leaf mold and/or wood chips.

Municipalities that maintain their own landfill will likely find this site suitable for a leaf composting operation. The availability of equipment and personnel at landfills could make leaf composting particularly practical. Leaf mold can also provide an excellent topping for finished landfills eliminating the need to obtain topsoil.

Composting Techniques

Leaf composting can be achieved either by heap or ditch composting or by windrow composting. Mechanical digestion is not a practical method for leaf composting because the time required to decompose leaves, even under ideal conditions, is relatively long due to the high C/N ratio of leaves. Heaps or ditches can be used effectively to compost leaves; however, windrows are usually preferred. Windrows allow efficient materials handling, provide good aeration, allow sufficient absorption of water, and can be easily formed by dumping truck loads of leaves in straight rows.

The preparation of leaves for good decomposition requires a few simple operations. Windrows should be shaped into uniform rows from 8 to 12 feet at the base and 8 to 10 feet high. Windrows built too high will have excessive compaction at the base resulting in anaerobic conditions. Windrows built too low may not allow sufficient insulation to sustain thermophilic temperatures during cold weather. Small windrows result in inefficient use of space. To insure proper aeration, it is important that tightly compacted clumps of leaves be broken apart. This can be achieved by fluffing the windrows using a front-end loader or using a tractor with a hydraulic bucket.

When the incoming leaves are not adequately moist, it is helpful to add moisture to the leaves. Adequate moisture (between 40% and 60%) is necessary for microbial growth. The addition of moisture also prevents wet leaves from blowing. Leaves can be mechanically reduced to break down their structure at the beginning of the composting process. This will induce more rapid decomposition but usually is not desirable because it increases the compaction making more frequent aeration necessary. Generally, preliminary mechanical reduction should not be practiced unless time and space limitations necessitate rapid decomposition.

Some publications indicate a belief that inocculation with a culture of microorganisms will enhance microbial activity and therefore increase the rate of decomposition. However, a large quantity and variety of microorganisms is present in leaves before the composting process begins. Because of the initial presence of a wide variety of microorganisms, it is questionable if the inocculation of additional microorganisms will have a major effect on the microflora found in the leaf windrows. Covering the leaves with 4 to 6 inches of finished compost (leaf mold) will supply a heavy inocculum of thermophylic microorganisms and also help prevent blowing of leaves.

To maintain aerobic conditions in a leaf windrow it is necessary to provide some form of aeration. Aeration is usually achieved by mechanically agitating the windrows with a front-end loader, a process usually referred to as turning. The major objectives of turning are to expose all of the material to air and to mix the materials from various parts of the windrow in order to enhance uniform decomposition. The fluffing action achieved by lifting the material and allowing it to drop through the air has been found to be effective. Another method of aeration that has been tried is the injection of compressed air into the windrows with long needles. The effectiveness of this method, however, has not been substantiated.

If infrequent turning is practiced, anaerobic conditions will occur in the center of the windrow after a period of decomposition. Decomposition in this area will slow down but the odors created by anaerobic decomposition should not be noticeable because they do not penetrate the outer portions of the windrow. The outer portions stay aerobic longer because of surface penetration of air and and an initially slower rate of decomposition. Odor problems can occur if accumulations of excessive amounts of water result in anaerobic conditions close to the surface.

The C/N ratio of leaves is relatively high. It can be as high as 80. This results in a slow rate of decomposition relative to that of other organic materials which have lower C/N ratios. The slow rate of decomposition enables leaves to be composted with relatively little maintenance. While a material such as horse manure needs to be turned every few days, leaves need to be turned only a few times a year and the exact time at which leaves are turned is not critical. Thus, leaf windrows can be turned during times when personnel and equipment are not needed for some other purpose.

The number and frequency of turnings required to complete decomposition depends on the composting time that is desired. If composting is to be completed in one year, at least three or four turnings will be necessary. Rapid decomposition is usually desired when space is limited or when there is a pressing need for leaf mold. Composting periods of more than one year are reasonable when space permits. In these cases less frequent turning will be required.

During extended periods without rainfall it may be helpful to add moisture to the windrows. This can be done most effectively during or right after turning when the water can penetrate throughout the windrow. During periods of moderate to heavy rainfall, water should not be added after the leaves are initially moistened.

Nitrogen can be added to leaves to increase the rate of decomposition. This increased rate of decomposition will result in greater oxygen utilization and in the need for more aeration. If a nitrogen source is to be added, caution should be taken to distribute the source evenly throughout the windrow. High concentrations of nitrogenous materials will likely result in undesirable anaerobic conditions. Sewage sludge, urea, and grass clippings are good sources of nitrogen. The addition of a source of nitrogen, such as dried sewage sludge, will have the beneficial effect of increasing the rate of decomposition and increasing the amount of nitrogen in the finished compost but it will likely result in an increase in costs due to the necessity of mixing the materials and providing more frequent turning. The pH of leaves will decrease during the initial stage of decomposition and then will gradually rise throughout decomposition to a final state which is near neutral. Lime, which has been used to enhance decomposition, is not needed. Leaf mold will have a final pH range of 6 to 7 and even leaf mold from oak leaves, which are initially more acidic than other leaves, will range near or above a pH of 6.

If environmental conditions are optimum, thermophilic temperatures will be reached inside the windrow and can be maintained for long periods of time. Cold weather can reduce the temperature in the areas of the windrow close to the surface but leaves are an excellent insulating material and allow high temperatures to be maintained within the windrow regardless of the ambient air temperature. A drop in the temperature is an indicator that either the environmental conditions are not adequate to support rapid thermophilic microbial growth or that the leaves have been decomposed beyond the point at which they can support rapid growth. Therefore, a drop in temperature will indicate that the environment in the windrow should be altered, which usually means aeration, or it indicates that the decomposition is nearing completion.

From the time leaves are collected to the time they are converted into leaf mold, they undergo a significant reduction in volume. Reports on volume reduction range from reduction of 8 to 1 to reductions of 30 to 1. The initial reduction in volume occurs when the leaves are collected. Most collection devices result in a compaction of between 3 to 1 and 5 to 1. Composting causes a volume reduction due to an increase in density and a loss of mass. Density increases naturally during decomposition as the leaves lose their structure and become more closely packed. Reduction of mass occurs during composting due to the oxidation of cellulose to form carbon dioxide and water. The final wet weight density of mechanically reduced leaf mold is reported to be approximately 1,500 pounds per cubic yard.

After decomposition is completed a number of finishing operations can be performed to further enhance the quality of the leaf mold. The degree of finishing practical will depend on how the leaf mold is to be used. If it is to be marketed, it may be practical to do a considerable amount of finishing but, if leaf mold is to be used as a final cover for a landfill site or similar bulk uses, it will need little or no finishing. Three likely objectives of finishing operations are: (1) removal of undesirable materials, (2) mechanical reduction, and (3) mixing leaf mold with soil.

Rocks, cans, glass, and other undesirable objects can be removed by screening. Screening can be accomplished with an inclined vibrating screen or with an inclined rotating cylindrical screen. Also, some mechanical reduction devices are capable of rejecting rocks and similar materials.

Mechanical reduction can be achieved with a number of different types of comminuters. Most commonly used are hammermills and shredders. Mechanical reduction improves the quality of leaf mold by reducing it into fine uniform particles. Leaf composting operations that practice mechanical reduction usually perform two reductions, one before the leaf mold is piled for storage and again before it is used. It is important to note that it is not necessary to do any finishing at all and, also, that leaf mold can be stockpiled for a long period of time to be further processed at a future date.

Mixing with soil can be done in conjunction with the mechanical reduction process. A mixture of 30 to 40% leaf mold with soil makes a topsoil-like material. Leaf mold is a humus material that can benefit soil which is deficient in organic material. Leaf mold can be used as a top dressing. It can be mixed with soil before or after application for use as topsoil, and it can be used as a mulch. Municipalities which are composting leaves are using leaf mold with good results on public lands. Leaf mold can also be utilized by the residents of a community for application to lawns and gardens. Some municipalities are selling their leaf compost to residents and commercial users while others are making it available free.

Tree Trimmings and Garden Wastes

Communities which have large quantities of leaves will undoubtedly also have large quantities of tree trimmings and garden wastes. Tree trimmings generated by municipal or private tree cutting crews are usually in the form of wood chips. In most communities garden wastes will be generated primarily at the residential level. Garden wastes consist largely of grass clippings, weeds, shrub and tree trimmings, and other plant residue. A leaf composting program could handle tree trimmings and garden wastes as well as leaves but these materials require special attention.

Wood has a high C/N ratio. Pine, for example, has a C/N ratio of over 700. A high C/N ratio together with a substantial physical structure make wood relatively slow to decompose. In order to compost wood in a reasonable period of time an additional source of nitrogen will be needed. This can be achieved by adding a concentrated source of nitrogen such as urea, sewage sludge, or grass clippings. Wood chips can also be mixed with leaves for composting. Wood chips added to leaf windrows will not decompose rapidly if in large concentrations, and therefore must be mixed quite thoroughly with the leaves.

Wood chips are also useful as a mulch and as a decorative ground cover. Wood chips used as a mulch for trees can hold in moisture and protect from extremes in temperature. The value of wood chips as decorative ground cover depends on their appearance. A chipper with sharp cutting blades and a properly adjusted clearance between the blades and bed plate will produce chips with a more desirable appearance. Also, the absence of small unchipped branches is preferable. Fresh wood chips will rob nitrogen from the soil because of their high C/N ratio. Therefore, wood chips can be used on areas where plant growth is not desired. Some communities that have leaf composting programs make wood chips as well as leaf mold available to the public.

Garden wastes are composed primarily of grass and weed clippings. These materials have C/N ratios around 20, and therefore decompose readily. Garden wastes can be composted with leaves but present a problem because they must be handled in the spring and summer as well as in the fall and because they decompose faster than leaves. Garden wastes need to be aerated more often than leaves because they utilize oxygen at a faster rate. If garden wastes are to be composted in the same windrow as leaves, they should be mixed thoroughly. This will reduce the variation in rates of decomposition in different sections of the windrow. A possible alternative is to mix garden wastes with wood chips thus increasing the rate of decomposition of the chips and decreasing the maintenance required to keep garden wastes aerobic.

Economics of Leaf Composting

The economics of leaf composting will vary greatly from one community to another. A representative cost figure for leaf composting would be misleading because costs vary considerably between communities depending on such factors as the availability of land, equipment, and personnel, the quantity of leaves to be handled, and the type of composting program.

Although it is not reasonable to assign cost figures to leaf composting that are representative of all communities, the costs for a particular community will likely give some indication of the general costs which are involved in leaf composting. Table 8.3 gives the cost of leaf collection and composting versus leaf collection and landfilling for Hinsdale, Illinois. The fact that it is a longer trip to the landfill site than to the compost site is a large factor making composting less expensive than landfilling in Hinsdale. A total of 570 additional man hours and 285 additional vehicle hours are required to haul the leaves to the landfill site as opposed to the compost site. Substantial savings are also realized by the elimination of the dumping fee and by the use of leaf mold to replace a portion of the needed topsoil.

There is an economy of scale with leaf composting which probably does not exist in landfilling or incineration. A large quantity of leaves should result in a lower cost per cubic yard of disposal through composting. The type of composting which is practiced also has a great effect on the cost. For example, a program which consists of forming windrows in the fall, turning them whenever manpower and equipment are available, and stockpiling whenever decomposition is completed will cost less than a program consisting of a full time employee responsible for the composting operation, specialized equipment such as mechanical reduction devices and mechanical screens, a planned schedule of turnings, the addition of urea, compressed air injection, etc. Most communities start leaf composting with a small program which may become larger and more elaborate with time. An important point is that a community need not go to great expense at first to try leaf composting.

Consideration should be given to the benefits derived from the use of leaf mold. Leaf mold can substitute for such materials as peat moss and topsoil. Most municipalities have a large need for these materials and the use of leaf mold can result in a significant financial savings. Some communities have had success in marketing leaf mold. Also the environmental benefit of improving the soil by replenishing the organic matter should not be overlooked.

TABLE 8.3: SUMMARY OF LEAF DISPOSAL COSTS – COMPOSTING VERSUS LANDFILL FOR HINSDALE

	Costs with Composting	Costs with Land Fill
Leaf Pickup (9,000 yds)	$26,309.93	$26,309.93
Dumping Costs (9,000 yds)		2,745.00
Additional Time to Landfill		
- 570 additional man hours		2,850.00
- 285 additional vehicle hours		3,825.00
Cost of Black Dirt		
- 330 yards	1,000.00	
- 1,000 yards		3,000.00
Turning and Shaping Windrows		
- 120 man hours	600.00	
- 120 hours front-end loader	420.00	
Shredding		
- 96 man hours	480.00	
- 48 hours front-end loader	168.00	
- 48 hours shredder	480.00	
Miscellaneous Expense	500.00	
Total Cost (9,000 yds)	$29,957.93	$38,729.93
Unit Cost (per yd)	3.33	4.30

ANIMAL WASTE COMPOSTING

In recent years, the tendency in livestock production has been toward increasing the number and density of animals on individual farms in order to maximize labor efficiency. This concentration of animal production in relatively small areas has resulted in a concentration of manure generation which, in turn, has created disposal and environmental problems. When animal manures are allowed to accumulate in open areas, a potential water pollution hazard exists. Rain water will leach organic matter and nutrients from manure. If storm runoff from areas of livestock production reaches a river or stream it can cause a serious pollution problem either by increasing the biological demand for oxygen or by providing nutrients which will stimulate an algal bloom.

Air pollution in the form of disagreeable odors can also be a serious problem. When animal manures decompose anaerobically, odors will invariably be produced. Anaerobic decompostion will occur when manures are stockpiled or when anaerobic liquid waste handling is used. The odors are sometimes detectable several miles away. The problem of odors has become more serious in many areas as urban sprawl has brought human residences closer to livestock farms.

Illinois is a major producer of livestock, and therefore is faced with handling large quantities of animal waste. In 1969 fifty percent of the agricultural income in Illinois was derived from the sale of livestock and of livestock products. Swine and beef cattle production make up the greatest part of the Illinois livestock industry. Illinois ranks second nationally in hog production and seventh in the number of feed cattle marketed. This large livestock industry in Illinois produces an equally sizeable amount of manure which must be handled in an environmentally satisfactory manner.

Composting is one of a number of alternative methods for animal manure treatment which are environmentally acceptable. There are a number of advantages that can be achieved through composting. One of the major advantages is that aerobic decompostion results in the elimination of the odors which are present when manure is allowed to decompose anaerobically. Furthermore, stockpiled manure compost will not emit odors even if stored in large quantities for long periods of time. This enables manure to be stored in an acceptable manner during periods when land application is not possible (the winter and periods of high rainfall).

Composting also eliminates the threat of serious odor production when the material is applied to the land. Flies can be controlled with the use of composting techniques since the thermophilic temperatures which are reached during composting will kill fly larvae. After aerobic decomposition is completed, the compost will not attract flies.

Another advantage of composting is the reduction of mass and volume which will facilitate handling and disposal. One experiment with thermophilic composting of dairy cow manure showed a 60% reduction of dry weight and up to 45% reduction in volume. Unless manure is mixed with a carbonaceous material, composting will cause a reduction of nitrogen which will be given off as ammonia. This decreases the nutrient value but increases the amount of material that can safely be applied to the land. Compost is also easier and more pleasant to work with than fresh manure. This increases its value as a soil conditioner for both large and small scale users.

Animal Waste Characteristics

The characteristics of manure will depend on the type of animal, quantity and quality of feed, housing, and waste handling techniques. Table 8.4 gives data showing variations between different animal manures. Manure from a particular animal will be a function of what that animal eats. If, for example, an animal is fed a fibrous material low in nitrogen, the feces will be mostly fibrous material and there will be little nitrogenous waste in the urine. On the other hand, if an animal is given a more than adequate supply of feed high in nitrogen, as is often the case with beef and poultry, the animal will utilize less and excrete more of the nitrogen. Housing and handling methods also have a great effect on the characteristics of manure.

TABLE 8.4: SUGGESTED VALUES FOR MANURE DEFECATION RATES PER 1,000 POUNDS LIVE WEIGHT IN CONFINEMENT ANIMAL PRODUCTION

	Dairy Cattle	Beef Cattle	Hens	Pigs	Sheep
Raw manure (RM), lb. per day	88	60	59	50	37
Total solids (TS), lb. per day	9	6	17.4	7.2	8.4
Total solids, percent RM	10	10	30	14.4	22.7
Volatile (VS), lb. per day	7.2	4.8	12.9	5.9	6.9
Volatile solids, percent TS	80	80	74	82	82
BOD_5, lb. per day	1.7	1.5	4.4	2.1	.7
BOD_5, lb. per lb. VS	.233	.252	.338	.363	.101
Nitrogen, percent TS	4	7.8	5.7	5.6	4.0
Phosphoric acid, percent TS	1.1	1.2	4.6	2.5	1.4
Potassium, percent TS	1.7	1.8	2.1	1.4	2.9

There are many different techniques of housing and waste handling used in the production of each type of livestock. If animals are allowed to graze on a sufficiently large area of land, their waste will not have to be collected and will usually not cause serious environmental problems. The trend in dairy, beef, and hog production is moving away from open grazing. When animals are confined to a relatively small earthen area, waste will accumulate and it will be in solid form since a portion of the moisture will either penetrate the soil or evaporate. When concrete flooring is provided, as under roofed areas or around feeding areas, the waste will be too liquid to handle as a solid unless bedding is provided to absorb the moisture.

Confinement housing has become more commonly used in livestock production. Poultry is virtually always raised in confinement housing while increasing numbers of cattle and hogs are being raised in total confinement. Some confinement waste management systems handle the wastes in a liquid state which requires the addition of water. In the case of bovine confinement systems, the waste will be handled as liquid if slotted floors are used and as solid if solid floors are used. Solid floors require the use of bedding material in order to absorb some of the moisture. Poultry confinement systems also either handle manure as a liquid by providing additional moisture or as a solid by allowing the manure to partially dry under the cages.

Moisture content is one of the important factors governing the compostability. The optimum moisture contents will be between 40 and 70% of manure, depending on the character of the material. Material containing a large percentage of carbonaceous bedding can be composted at higher moisture levels, whereas wastes high in nitrogen, such as poultry manure, compost more effectively with lower moisture contents. Liquid manure handling systems obviously do not lend themselves well to composting. Manure will seldom be initially too dry to be effectively composted but in most cases it will be adequately or more than adequately moist. Generally, those systems which provide some form of moisture absorber such as mixture with bedding and/or soil penetration, will provide manure which has a moisture content adequate for composting.

The C/N ratio of animal excrement will usually be lower than 20 and can be lower than 10. Poultry and cattle manures are particularly high in nitrogen content. A low C/N ratio indicates that there is more nitrogen than can be used by the microorganisms during digestion of the organic matter. Nitrogen exists primarily as urea, as undigested protein, and as microbial tissue. A portion of the urea will combine with water to form ammonium carbonate. Ammonium carbonate is unstable and can undergo decomposition to form ammonia, carbon dioxide and water.

Ammonium carbonate will decompose more readily under elevated temperatures, such as is the case with thermophilic composting. Also, pH values above 7 will decrease the stability of ammonium carbonate. Ammonia in the compost pile or windrow can also be converted into nitrate by nitrifying bacteria, and nitrate can be converted to nitrogen gas through denitrification. However, nitrifying bacteria do not live at thermophilic temperatures and therefore this reaction is limited to the surface of the compost pile. Denitrification occurs only under anaerobic conditions; thus, nitrogen is given off primarily as ammonia through volatization and, to a limited extent, as nitrogen gas through denitrification. A considerable reduction in nitrogen can be expected if the initial C/N ratio is low.

The addition of a carbonaceous material such as straw, wood bark, saw dust, or corncobs can often help the composting characteristics of a manure. Carbonaceous materials can be used to reduce moisture content in manure and to raise the C/N ratio to a more desirable level. Table 8.5 gives the example of fresh poultry manure mixed with corncob. In this case the C/N ratio was raised to a desirable 29/1 ratio and the moisture content was reduced to 51.9%, which is also a desirable level.

TABLE 8.5: IMPROVEMENT OF COMPOSTING CHARACTERISTICS OF POULTRY MANURE BY MIXING WITH CORNCOBS

	Fresh Manure [A]	Ground Corncob [B]	Mixture [2A+B]
Percent Moisture	75.4	5	51.9
Percent Nitrogen	1.13	.25	.83
Percent Carbon	8.0	51.8	24.5
C/N Ratio	7/1	206/1	29/1

Composting Techniques and Examples

Rotary drum mechanical digesters have been used to compost poultry manure. These worked best when a dry carbonaceous material was added to improve structure, moisture content, and C/N ratio. Generally, mechanical digesters will be impractical because of the capital investment that is required and because there will normally be space available for windrow composting.

One composting experiment was carried out with dairy manure using bins in which air was introduced into the manure with low pressure blowers which discharged air through perforated pipes in slabs on the bottom of the bins. The director of the experiment, Charles L. Senn, believes that this method is an efficient and economical method of composting, stabilizing and drying cattle waste. The results obtained in this experiment were:

(1) a moisture content of 40 to 60% was most satisfactory;
(2) a mixture of 10% of finished compost, or more, mixed with the raw manure, appeared to improve the end product;
(3) 4 to 7 days of aerobic composting in each of two bins, plus 30 days' storage for aging, produced a product with the desired qualities;
(4) air flow rates began at 3 to 5 cubic feet per minute per cubic yard at the start and were gradually reduced, as needed, to maintain temperatures at over 140°F but below 174°F;
(5) air pressures averaged 4 inches of water gauge (0.15 psi) at the start and were decreased to 1 inch water gauge by the end of the aerating period.

It was also concluded that windrows could effectively be used instead of bins. This would lower construction costs and allow the manure to be handled with tractor and truck, which will normally be available on a livestock farm.

Windrow composting has been successfully utilized to compost manure. Although it has been most frequently used to compost poultry manure, it has also been used with wastes from other animals. The following is a quotation from an article written by John S. Wiley which describes the Sunrise Compost Company, which was composting feedlot and meat packing wastes at Parker, Washington.

> Mr. Ted Williams operates a windrow composting plant that treats a mixture of feedlot manure from 5,500 steers and meat packing wastes, including blood. The windrows are about five feet high and contain approximately one ton of wastes per one and one-half feet of length. Windrows are turned six times during the six-week composting period, using a front-end loader and a large truck-mounted manure spreader. The loader used is a Caterpillar 933 Traxcavator with a one and one-half cubic yard bucket.
>
> In forming the windrows, Mr. Williams first makes a windrow

by mixing steer manure and paunch manure from the packing plant. The steer manure has generally been stockpiled for about six weeks. A trough is made along the top of the windrow to receive blood from a tank truck. The wastes are then put through the spreader which mixes them and forms new windrows. Mr. Williams reported only a minor fly problem with spray required only once during 1963. All wastes except horns, ears, and tails are composted. These wastes are dumped into a pit and buried in a landfill.

The compost is then processed through the following equipment, which was homemade:

(1) A ¾-inch vibrating screen (tailings are mainly sticks and stones and are discarded).
(2) A ³⁄₁₆-inch rotary drum screen with interior angle-iron lifters.
(3) A miller-type impact mill where material retained on the drum screen is ground.
(4) A current, single-pass, rotary drum dryer.

Part of the compost is stored in bulk and part is bagged. The compost is sold as Wake-Up Soil Conditioner and Composted Steer Manure mostly in the Seattle, Portland and Spokane areas.

In a report prepared by John R. Snell Engineers, Inc. of Lansing, Michigan at the request of Cobey Environmental Controls Company, Inc., an experiment in composting feedlot manure with the use of a Terex-Cobey Composter was described. In this experiment windrows were turned every day using the Terex-Cobey Composter. The manure consisted of 50 to 60% urine and fecal matter and 40 to 50% bedding which was ground bark, sawdust, and wood chips. This material, which had been standing for 4 to 8 weeks in covered feedlots, was loaded onto trucks with a front-end loader. The trucks unloaded the material to form the initial windrows. Much of the material was hard pack lumps, 6 to 12 inches in diameter. The turning device weighs 53,000 lb and operates at 1 to 1½ miles per hour, and is capable of turning 800 to 1,000 tons of material per hour.

In this experiment, the moisture content was maintained around 40%. When the percent of moisture fell below 40%, a fall in temperature was also observed signifying a reduction in rate of decomposition. It was suggested that the moisture content be kept above 40% and that 50 to 55% is the optimal moisture content. Frequent turning accelerated the rate of moisture reduction. If insignificant amounts of rainfall are experienced, moisture should be added to the windrows to maintain adequate moisture content.

The pH of the manure was initially 8.3 and finally 9.1 to 9.4. The high pH was caused by the urine content of the mixture. Despite this high pH the material appeared to compost readily. This report indicated that feedlot manure could be composted in less than three weeks with the use of the Terex-Cobey Composter.

Another report describing the use of a Terex-Cobey Composter was prepared by Woodward-Envicon Inc. of San Diego, California. In this experiment dairy manure from the Alta-Dena Dairy, Chino, California was composted. The manure was collected from feeding corrals where no bedding was used. The average age of manure was approximately two months. The manure was composted in windrows for four weeks with seven turnings during the first two weeks and six turnings during the last two weeks. Table 8.6 describes the change in composition that resulted from composting.

TABLE 8.6: CHANGE IN CHARACTERISTICS OF DAIRY MANURE DURING COMPOSTING

	0 Days	14 Days	28 Days
Percent volatile organic materials	42	36	27
Percent Carbon	23	18	14
Percent Nitrogen	1.3	1.1	0.9
Percent Phosphorus	--	--	.4
Percent Potassium	--	--	2.0
Percent Moisture	40	35	49
Volume, Cubic Yards	650	--	580

An important factor in composting poultry manure is keeping the droppings under the cages as dry as possible. Maintaining the watering systems so they do not leak additional moisture onto the manure will help. Also a ventilation system to provide air movement in the cage houses will help to keep the under cage droppings dry. Pennsylvania State University has been experimenting with a system which mechanically stirs the droppings. This stirring aids in drying and the aeration it provides allows aerobic decomposition to occur at a rapid rate. If the manure is kept dry enough under the cages, anaerobic conditions can be avoided and composting can begin before the manure is removed from the houses.

Deep-pit or high-rise poultry housing have been used to compost manure without removing it from the houses. In these systems the manure is kept dry and adequate ventilation is provided in order to maintain aerobic conditions. Manure can be allowed to accumulate for several years in these houses if proper conditions are maintained. If the moisture content is not too high (60% or less), poultry manure can go directly to the windrow. If the manure is too moist, it should be mixed with a dry carbonaceous waste such as sawdust, rice hulls, or corncobs. Windrows should be located on the poultry ranch to minimize transportation. According to one source, turning is not necessary more than once a week but at least three turnings are required to obtain a reasonable level of stability.

Little work has been reported on composting hog manure. It is assumed, however, that the same factors hold for hog manure as hold for other manures, and therefore that hog manure could be successfully composted.

If the compost is to be used on agricultural land, little or no finishing should be required. Composted manure will be more easily handled than uncomposted manure and can be applied with traditional manure spreading equipment. If the compost is to be bagged for marketing to small users, grinding and/or screening will help appearance and salability of the product.

COMPOSTING INDUSTRIAL ORGANIC WASTES

The one major application of the composting process which has been the most underutilized in the United States is for the disposal of organic industrial wastes. Several European industries make extensive use of composting for decomposition of their waste organic residues but, with the exception of the pharmaceutical industry, few industries in the U.S. have even seriously considered composting. There is a great potential for the industrial use of composting particularly for the decomposition of materials which break down slowly in biological wastewater treatment processes.

The types of industrial wastes which have been successfully handled by composting in Europe are listed in Table 8.7. It should be noted that materials like pulp, hair, connective tissues of animals, cellulose, fibers, oil sludges, etc., which do not normally decompose in biological wastewater treatment processes and often cause clogging problems with screens, pumps, and air diffusers, have been successfully composted. Most of these materials are of animal or plant origin (even crude oil) and are decomposed by microorganisms; however, they break down slowly and, due to their physical characteristics, are not easily handled by the mechanical devices which are employed for biological wastewater treatment processes. They may be more easily handled as solids and readily broken down during a composting process if the environmental conditions for the process are maintained properly.

The food processing industry is probably the largest source of industrial organic wastes in Illinois. There are over 110 food processing concerns in Illinois which have a-net worth of over one million dollars. Some of the other industries in Illinois which produce large quantities of organic wastes are the pharmaceutical, petroleum, paperboard, and leather industries.

All industries producing organic wastes have five alternative methods for disposal; namely, (1) utilization for by-product production, (2) dilution with water for biological wastewater treatment, (3) drying and incineration, (4) landfill, and

(5) composting, either alone or after mixture with other organic wastes. By-product production is always the choice alternative especially if the by-product can be sold for a profit. However, most industries already use this method to the extent that it is feasible for economic reasons.

TABLE 8.7: INDUSTRIAL ORGANIC WASTES SUITABLE FOR COMPOSTING

Industry	Type of Waste
Food Industries	
Sugar Mills	Pulp and Sludges
Starch Factories	Pulp and Fibers
Canning Industry	Refuse of Fruits and Vegetables
Chocolate Factories	Cocoa, Nutshells, Fats
Breweries	Spent Mash, Filtrate Residue
Distilleries	Hop Waste, Press Residue
Meat Packing Plants	Bones, Horns, Hoofs, Paunch Manure, Meat Scraps
Pectin Factories	Fruit Pulp, Filtrate Residue
Wood Products Industries	
Pulp and Paper Mills	Bark, Spent Liquors
Saw Mills	Sawdust, Chips, Shavings
Textile Industries	Wool, Hair, Cellulose Fiber, Waxes, Viscose Residue
Leather and Glue Industry	Hair, Skin, Fats
Pharmaceutical Industry	Fermentation Tank Sludge Filtrate Residues
Petroleum Industry	Crude Oil, Sludges

Biological wastewater treatment processes have been developed to a high degree of efficiency for industrial application over the past twenty-five years and thus have been widely utilized. However, they do have their limitations and, since the adoption of new effluent standards and water quality standards by the Illinois Pollution Control Board in the spring of 1972, many industries are finding that these processes are not entirely adequate to achieve the degree of treatment required.

Tertiary wastewater treatment processes; such as lagooning, sand filtration, lime precipitation, etc., may be added-on after the biological wastewater treatment processes in order to meet the water quality regulations. However, in many cases it may be more economical to reduce the load on the wastewater treatment process by removing and disposing of some of the organic waste as a solid and thereby achieve the effluent quality required.

Drying and incineration as a method for disposal of organic wastes is practiced by some industries, particularly in those cases in which the waste can be added to coal or oil used to fire a boiler. This technique not only disposes of the waste but also recovers some heat value. Drying is such an expensive process and incineration is so difficult to control that this technique will probably not be used except in those cases where some heat value can be recovered.

Landfill is presently a popular method for disposal of industrial solid waste in Illinois. It is quite economical when convenient landfill sites are available and can be inoffensive if properly done. Suitable landfill sites in Illinois are rapidly being used up and, in many cases, industries will have to switch to other methods of disposal or face rapidly increasing transportation costs.

When applicable, composting will have several advantages for disposal of organic industrial waste. Although it requires extended periods of time in comparison with other disposal methods it does not require as large a capital investment to get started. Other than provision of the land area required, most of the costs of composting can be written off as operating expenses. If properly done it is inoffensive and can handle resistant organic matter which causes problems with the alternative disposal methods. A large reduction in the solid waste volume is achieved by composting. The disposal of the finished compost can contribute to improving soil quality and does not contribute to air or water pollution. Some industries might want to consider composting for waste disposal primarily for the public relations value of being able to say that their waste disposal techniques are ecologically sound.

Requirements

The major difficulty with composting most organic industrial wastes is that the waste is too homogeneous. To achieve a reasonable rate of composting the organic material must be in particles small enough to have a large surface area and large enough so that air can move freely between the particles. The material must also be wettable and, at the same time, drain off excessive moisture readily. It also needs to have the proper balance of the required nutritional elements. Most organic industrial wastes are either too fine (sawdust), too coarse (packing house offal), nonwettable (oils and waxes), retain too much moisture (fruit pulp), have too much nitrogen (paunch manure), or too little nitrogen (cellulose fibers) to compost readily alone. Careful attention must be given to the preparation of the waste for composting, to nutrient supplementation, and to mixing different types of waste together in order to obtain a material which is readily compostable.

The requirements for rapid composting of organic industrial wastes are the general requirements for composting. The proper microorganisms must be present in waste having the proper chemical composition and physical structure and they must be supplied with the proper amount of moisture and oxygen.

Unlike municipal refuse, leaves, and animal manures, organic industrial wastes may not initially contain enough of the types of bacteria and fungi required for rapid decomposition. This is entirely a start-up problem. Once a batch of waste has been successfully composted the finished compost will contain myriads of the spores and other inactive stages of the microorganisms which decomposed the organic matter. Merely mixing a small amount of the finished compost with the next batch of waste will supply an adequate population of microorganisms.

The initial start-up problem may be overcome using readily available materials. Garden soil contains an adequate population of microorganisms to decompose most wastes and, when mixed with the raw waste at a proportion of 1 part soil to 100 parts of waste, will provide the required bacteria and fungi. Domestic wastewaters also contain a large variety of microorganisms and settled sewage from a municipal wastewater treatment plant may be used both to start the culture and to supply moisture if needed.

The organic waste should have a physical structure which will produce a compost heap having a loose or porous texture. Large pieces of organic wastes can be reduced to the proper size with shredding or grinding equipment. However, many industrial wastes have too small rather than too large a particle size. This problem can be overcome by mixing the waste with solid waste from another source, such as municipal refuse.

Industrial wastes are not likely to have nutrient compositions ideal for microbial growth. It is important to note, however, that the composition does not have to be exactly ideal for effective composting. For example, a C/N ratio of 30/1 is considered to be ideal but ratios as low as 10/1 and as high as 50/1 can also be satisfactorily composted. Industrial wastes may have disproportionate amounts of nutrients other than carbon and nitrogen, such as phosphorus or potassium. Nutrient ratios can often be adjusted by mixing wastes of different nutrient compostions. When this is not practical, chemical fertilizers can be added to supply nutrients not present in sufficient quantities.

Wastes from some industries may have extremes in acidity or alkalinity. Microorganisms grow best under neutral conditions; however, slow decompostion will take place at high or low pH values. If the materials are only slightly acidic or alkaline, pH control will not be necessary and will likely be impractical. However, in extreme cases, pH control can be helpful.

Maintenance of adequate supplies of oxygen and water should be provided by the composting process. Many organic wastes will contain excessive amounts of moisture and will require a reduction of water concentration before composting. Mixing with another organic waste that is low in moisture is one solution. There are also numerous processes which mechanically extract moisture and could be used to obtain a reasonable water concentration. Water concentrations of 40 to 60% are generally considered to be desirable. Materials with water concentrations not too much greater than 60% could be effectively composted if adequate aeration were supplied. Some degree of dewatering can be achieved with the compost process itself.

Perhaps the most useful technique in preparing organic industrial wastes for composting is mixing wastes of different characteristics. Wastes with high and low C/N ratios, high and low pH's, or fine and coarse structure can be mixed to create a material with better composting properties. This technique is particularly attractive when a suitable waste is readily available.

Examples

Generally, if an organic industrial waste is in a favorable state for composting or can be altered to reach a favorable state, it can be composted with existing techniques.

Example 1: Pharmaceutical Waste – Lederle Laboratories at Pearl River, New York, carried out a literature search, a pilot study, and field experiments to determine the best method of composting sludge from their waste treatment plant. The resulting process was used to compost all the sludge cake, 6 to 10 tons per day, from their waste treatment plant. The sludge originated from fermentation of antibiotics, extraction of pharmaceuticals from animal livers, and the sanitary sewage of 4,000 employees. Composting was considered the best and most practical method of treatment, disposal, and utilization of these wastes.

The sludge was dewatered in vacuum filters. Lime was used to aid the dewatering process. The resulting sludge cake had a water concentration between 65 and 85%, a pH between 7.5 and 8, and a C/N ratio of 15/1. Pilot plant studies showed that a mixture of 65% sludge cake, 25% stable manures, 10% sawdust, and .05% phosphate rock met both the composting and materials handling requirements. The mixture contained 60% water and, after one day of composting in the pilot study, had a pH of 6.3 to 7.5.

The pilot plant study showed that the mixture could be composted in six days but this required the maintenance of optimum conditions. In actual practice, a much longer 150-day process was used because adequate space was available and the longer process proved more economical. The equipment used for the large scale process was a one cubic yard capacity front-end loader and a 30-cubic yard per hour capacity shredder.

The materials were mixed by dumping the sludge cake on a prepared bed of sawdust and animal manure and by picking up the material from the outer edges of the pile with the front-end loader and making a new cone-shaped pile. One part of finely ground rock phosphate was added for each 200 parts of mixture. Seeding was also practiced by mixing in some previously composted material. The blending process was completed by shredding the mixture with a rotating belt shredder.

The shredded material was piled in large windrows 25 feet wide, 15 feet high, and 300 feet long. The materials were composted in the windrows for five months. No mechanical aeration was provided so the decomposition process was partially anaerobic. After the five month period the material, mostly large clumps, was reshaped into cone-shaped piles to permit additional drying. After a drying period the compost was reshredded, yielding a granular, free-flowing, dark brown material. The final product was used on the grounds of the pharmaceutical plant or by the nearby communities or was marketed through a distributor mainly for lawn treatment and soil conditioning.

Example 2: Fruit Wastes – An extensive study was made by the National Canners Association on composting fruit and vegetable refuse. Although some composting was done with vegetable wastes, most work was done with fruit wastes. Three different methods of composting were used. In the first method, wood bins with a one cubic yard capacity were used. Turning and aeration were achieved by manually turning with a shovel. This first method was used to obtain information for later, large scale experiments. The second method used was open windrows. These windrows were placed on an asphalt surface and were turned and aerated each day by a modification of a machine designed to remove rock and gravel from quarries.

The third method used was closed-wall windrows. Pairs of walls which were six feet high and forty feet long were constructed nine feet apart on the asphalt surface. Forced air aeration was supplied through the floor of the windrow and mechanical turning was provided. The mechanical turner used for the open windrows was used at first, but later a specially designed turner was built that ran on tracks on the walls of the closed-wall windrows.

The solid fruit wastes were ground producing a pulpy slurry with moisture levels around 90%. The liquid nature of this material necessitated the mixing with a moisture absorbing material. Rice hulls, sawdust, coffee ground residue, shredded redwood bark and wheat straw were used to absorb moisture. Of these, shredded redwood bark and wheat straw were found to be inadequate, sawdust and coffee ground residue were found to be adequate, and rice hulls were found to be the best material tried. After the mixture of fruit wastes and rice hulls attained thermophilic temperatures, periodic additions of fruit wastes were made to the windrows. This greatly improved the waste handling capacity of the windrows.

Example 3: Spent Sulfite Liquor – Experiments in composting spent sulfite liquor were carried out in Switzerland. Spent sulfite liquor is a waste product of the cellulose extraction process. It is a common waste product from the paper, textile, food, pharmaceutical, and fermentation industries. This material is particularly difficult to treat with traditional wastewater treatment processes, and on first assessment it appears that it would be difficult to compost due to its physical and chemical characteristics. It is quite acidic (pH 3), has a high C/N ratio (over 300/1), and a high water concentration (88 to 92%).

In this experiment the spent sulfite liquor was mixed with ground municipal refuse at two different concentrations. The first was 695 pounds (83.2 gallons) of sulfite liquor per ton of refuse. The resulting water concentration was 43%, and the pH was 8.4. The mixture and a control of 100% refuse were composted in piles 4.2 feet high. The mixture composted well and tests with wheat showed the compost from the mixture to be as good if not better than the compost made from ground refuse alone for disposal on soil. The second concentration was 1,320 pounds (158 gallons) per ton of ground refuse. This mixture was composted in laboratory fermentation cells (perforated plastic baskets). The water concentration of the mixture was 75%. This high a moisture level can result in anaerobic conditions until a portion of the moisture is lost. The second concentration of spent sulfite liquor also proved to compost well and the final compost was at least as good as pure refuse compost.

Organic waste other than municipal refuse could also be used to create a compostable mixture. The requirements of the other wastes would be a relatively low concentration of water and a relatively high concentration of nitrogen and other nutrients. A good sawdust-tannery waste mixture, for example, could prove to be a good mixture to add to spent sulfite liquor.

Summary: These examples demonstrate that many organic industrial wastes can be readily decomposed by composting. The composting process greatly decreases the volume and weight of the waste material and converts it into an inoffensive product which is easily mixed with soil. Experience with composting organic industrial wastes has been quite limited in the United States. Industries considering composting should approach the process on an experimental basis and expect to spend several months on pilot plant work before developing an entirely suitable process.

UTILIZATION OF COMPOSTS

The obvious and traditional use of compost is as a soil amendment. The microbial activities which are responsible for the changes occuring during composting continue when organic matter is mixed with soil but at a much lower rate. If raw organic matter were to be applied in large amounts to soil, the soil would become sour (anaerobic) and would not be suitable for agricultural or garden use until the period of rapid decomposition of organic matter was completed.

In this context, composting can be considered to be a technique for accomplishing the rapid decomposition of organic matter in an aerobic environment so that the organic matter can be applied to the soil for the very slow final decomposition. The agricultural areas of the world could easily accept all of man's organic wastes and benefit from it, if the waste were first properly prepared by composting. The uses of compost other than for improving the quality of soil are quite minor. The only one considered in this chapter is for mushroom production.

Effects on Soil

Compost can benefit soil by supplying plant nutrients and by supplying humus. Although the addition of plant nutrients is an important effect of compost, it is usually secondary to the effect that the addition of humus has on the soil. Humus has been defined as a complex mixture of amorphous and colloidal substances that develop naturally as raw organic matter is digested by various microorganisms. It is a natural material, though exceedingly variable and heterogeneous, and it possesses properties that distinguish it sharply from the original parent tissues and from the simple products which develop during its synthesis. Humus has an important effect on both the chemical and physical properties of soil and therefore is a major factor for determining soil fertility.

The inorganic part of the soil is classified as three materials depending upon the particle size: clay particles have a diameter of less than 0.002 mm; silt particles have a size range of 0.002 mm to 0.05 mm; and sand particles have a size range of 0.05 mm to 2.0 mm. Any particles larger than 2 mm in diameter are considered to be gravel and rocks. The U.S. Department of Agriculture uses a three-way classification of soil based on the percentages of sand, silt, and clay.

Although the physical skeleton of soil is made up of decomposed particles of rock, the properties of the soil are largely dependent upon the chemical nature of the soil and the amounts of moisture and organic matter present. The chemical nature of the soil can be described by its exchange capacity and its degree of aggregation. Soil particles have reactive chemical groups (exchange sites) on their surfaces. The smaller the particles the greater the surface area per unit weight of soil and the greater the exchange capacity.

Thus, clays have much greater exchange capacity than silts or sands. Metallic ions such as sodium, calcium, iron, aluminum, magnesium, etc., are bound to the exchange sites. The amounts of the different metallic ions can be changed by the material dissolved in the water percolating through the soil. For instance, if water with a low pH percolates through the soil, hydrogen ions may replace the metallic ions at the exchange sites producing an acid soil.

The small clay and silt particles aggregate together forming large clumps. The tendency of the soil particles to aggregate depends to a large extent upon the type of metallic ions on the exchange sites. Clay particles saturated with sodium have a tendency to swell and aggregate into a sticky mass. Calcium or hydrogen ions on the exchange sites produce a soil of a finer texture. Ideally, a soil should have enough aggregation so that it works well but not so much that it becomes sticky and impenetrable to water.

Humus can change the physical nature of soil far in excess of its proportional content. The most apparent visual effect of humus is to darken the color of soil. More important is the effect that humus has on soil structure. Humus can greatly improve soil structure by increasing aggregation and protecting aggregates from destruction by water, thus increasing tractability, porosity, and aeration. The changes in structure have a net effect of making soil easier to manage and more suitable for plant growth.

All soils have a porosity (void space) in a limited range of approximately 40 to 60%. This space between the soil particles may be occupied either by water or by air. During periods of rainfall the surface water infiltrates into the pore spaces in the soil and may completely saturate the soil by driving out and absorbing the air. After saturation with water, sandy soils drain readily because of the large pore size and may dry out quickly. On the other hand, clay soils retain moisture because the pore openings between particles are so small that the water is held in the pores by surface tension. In order for clay soils to drain the very small soil particles must be aggregated into larger clumps which have large pores between them. The roots of plants need to draw both water and oxygen from the pore spaces in the soil. If a soil stays saturated with moisture for too long a period, it may become anaerobic and unsuitable for root growth.

The colloidal nature of humus increases the capability of soil to absorb and retain water and the aggregation of small soil particles caused by humus increases the percolation of water through the soil. The result is increased moisture uptake from the soil which increases availability of water to plants and decreases runoff and erosion.

Humus is composed primarily of materials such as oil, fats, resins, polyuronides, and lignins which decompose very slowly. It is believed that these materials are modified during decomposition in such a way as to enable them to unite with proteins and other nitrogenous compounds. The result is an increase in the availability of nitrogen, phosphorus, sulfur, and other nutrients to plants growing in the soil. The cation exchange capacity of humic material also increases the availability of nutrients. Humic particles can attract nutrients and hold them strongly enough so they will not be leached away but loosely enough so that they are available to plants. Humus also increases the availability of phosphorus by attracting iron and aluminum ions of phosphate salts, thus setting fixed phosphorus free.

The type of effect the humus has on soils varies to some extent with the type of soils. Sands and soils with coarse textures allow rapid water percolation but have poor water absorbing and retention characteristics. Humus can greatly increase the water holding capacity of coarse texture soils. Humus also improves the structure of sands by giving them a needed cohesiveness. Perhaps the most important beneficial effect that humus has on sandy soils is to improve its nutrient holding characteristics.

Fine particle clay soils have good water and nutrient holding capacity but these heavy soils have a plastic and sticky texture that is difficult to work and that does not allow adequate percolation. Humus encourages granulation in these soils which improves texture and allows percolation of water through the surface soil. Humus will also increase the water and nutrient holding capacity of these soils. The properties of loams, though superior to either sands or clays, can also be helped by the addition of humus, especially by increasing water and nutrient holding capacities.

Mushroom Industry

The mushroom industry has been practicing composting for years since the compost is used as a culture on which mushrooms are grown. Horse manure and straw are common ingredients in mushroom compost. Gypsum and dry malt may also be added to improve the mushroom growing characteristics of the compost. The mushroom farms in the Chicago area compost horse manure from nearby race tracks, which often have an adequate amount of straw already mixed in.

Mushroom growers utilize the windrow method of composting. The combination of proper environmental conditions maintained by the mushroom grower and an initial material which is well suited for composting results in a very rapid decomposition. Many growers utilize mechanical windrow turners which have been designed especially for use in the mushroom industry. Moisture is added if needed during the turning operation. The composting process is usually completed in 10 to 11 days with 3 or 4 turnings during this period. Thermophilic temperatures which are reached during the first or second day of composting are especially important to mushroom growth because they kill organisms that could cause disease in mushrooms.

After the composting process is complete, the compost is loaded into shelves in mushroom growing houses where it is seeded with mushroom spawn. Temperature and atmospheric moisture are maintained at constant levels in the mushroom growing houses. The compost will start producing mushrooms after approximately 30 days and will continue to produce for several months. After the compost becomes contaminated with organisms harmful to mushrooms or after its rate of mushroom production drops significantly, it is removed and marketed to the public as a soil conditioner.

Inner City Plant of Ecology, Inc.

One obvious problem facing municipal solid waste disposal is that the spatial distribution pattern of generation is nearly inverse to that of the most cost effective existing techniques for disposal. This suggests widespread, intensive reliance on transportation or on the potential of development of new technology. PB 230,140, 1973, prepared by B.W. Vincent and J.A. Ruf of Ecology, Inc. describes one attempt at development of a technological approach. The attempt is to develop the economic viability of composting municipal refuse in an inner city situation.

The sources of information were site visits and interviews with officials of the developing company, accountants' reports supplied by the company, and interviews with and reports by officials in New York City Environmental Protection Administration.

The plant closed about June, 1973, during the preparation of the report. It was closed because of lack of funds; general reorganization of the company; including new refinancing, to allow the construction of larger plants; and need to modify equipment in order to be self-energizing. The plant is scheduled to reopen about June, 1974.

SUMMARY

The composting facility owned and operated by Ecology, Inc. in Brooklyn, New York, appears to offer an environmentally sound method of disposal of municipal solid wastes. The facility employs a process which subjects raw municipal refuse to environmental conditions suitable for biological activity in such a way that natural degradation processes are accelerated, achieving effective stability within forty hours. The remains after this activity appear to be environmentally inoffensive. The cost of this process per incoming ton of raw refuse for a 600 ton per day plant is estimated by Ecology, Inc. as $5.34 operating expenses plus capital expenses of $5.80 per rated ton of capacity, amortized for 20 years at 6% plus land costs. These data indicate that the total cost per ton of refuse for the conversion process in New York City, for a 600 to 1,000 ton rated capacity, could be less than $15.00; according to city auditor figures, this is competitive with new or upgraded incineration facilities.

Bulky and large ferrous wastes are removed as usual in incineration; there is little unconverted residue comparable to incinerator ash; there is no reported air pollution, the major gaseous emissions are water vapor and carbon dioxide; there is no water pollution comparable to quench water; and the versatility of the product is much greater.

It still retains two thirds of its raw Btu content for subsequent use in energy recovery, and it may be used as a fertilizer carrier, a building aggregate or other saleable raw material. The cost of processing the conversion product for sale as fertilizer will permit a considerable return from sale in subsidy to the whole operation. In the estimation of Ecology, Inc. it will render a net profit as well. Processing for sale as fuel or construction material are less profitable but still self-supporting.

The refuse conversion process in operation at the Ecology, Inc. facility in Brooklyn offers considerable promise as a candidate system for inner-city disposal of domestic solid waste. This conclusion was based on two assumptions: (1) that intracity disposal is in fact desirable, as opposed to regional systems with transhipment and, (2) that the cost estimations of interviewed officials are in fact valid. It is also based upon casual observation of the system, its apparent environmental impact, its product, and on independent analyses of the product. The conclusion of this investigation are as follows:

(a) The process receives municipal domestic refuse as collected, removes 15% by weight as bulky rejects and tramp iron salvage, and processes the remaining 85% to an environmentally inoffensive product in a manner compatible with urban industrial land use.

(b) The product of this process is useable as a carrier for fertilizer; it also lends itself to separation into fractions which may prove useable as construction materials and as fuel.

(c) The economic viability of the conversion process as presented by the company appears plausible, and the addition of chemical nutrients to the product for sale as fertilizer seems economically sound. (It must be borne in mind, however, that company cost and market projections were used without attempt at validation).

(d) The process is capable of handling domestic refuse without apparent environmental degradation, and subsequent reuse of its product is no more environmentally deleterious than is the use of the materials for which it may substitute.

PRODUCTS

Ecology, Inc. intention is to produce saleable products which use domestic municipal refuse as a raw product. The intended market presently is the retailing of fertilizer, fuel, and construction materials. Refuse is received and rendered inoffensive by stimulating biochemical action, producing Stabilate, which is used as a vehicle for chemical nutrients, as a fuel, or as a fine aggregate for construction materials. This Stabilate is classified into five categories by the company according to its characteristics and intended use. Some characteristics are given in Table 9.1.

TABLE 9.1: STABILATE CHARACTERISTICS

Stabilate Grade	Bulk Density (lb./cu. ft.)	Moisture	% Retained on Sieve No.					
			6	8	10	12	20	pan
I	15-20	4-6	15	5	10	40	10	20
II	30-40	8-10	15	10	5	60	5	5
III	8-12	3-4	30	35	5	5	5	10
IV	To be determined for product specifications.							
V	Dependent upon IV. (V=III-IV)							

Results of the calorimetic tests run on Stabilate III by New York University's Civil Engineering Department are as follows:

BOMB CALORIMETER

Gram Sample	Charred Residue	Btu Per Pound	Moisture
.0556	.0245	7566	7%
.2470	.0538	7500	7%
.4065	.0890	7725	7%

BARNETT TEST

Washed Sample:	7841 Btu per pound
Ash Weight:	0.0897
Sample Weight:	0.0178
Whatman Filter Paper:	7,445 Btu per pound

(continued)

TABLE 9.1: (continued)

OTHER CHARACTERISTICS	
Specific Gravity:	1.66
Moisture:	6.5%
Fiber Length:	0.6 to 1.0 mm average 20 to 35% is less than 0.2 mm 20 to 40% is between 1.0 mm and 3.0 mm

Generally, Stabilate I is the untreated product of the digester and may be used directly as a fertilizer carrier or may be processed into Stabilate II or III. Stabilate II may prove to be useful as an aggregate in bricks or blocks. Stabilate III is used as fuel or soil conditioner or may be separated into Stabilate IV for heavy fiberboard, and Stabilate V for light. See the Flow Diagram in Figure 9.1.

FIGURE 9.1: MATERIALS FLOW BY WEIGHT

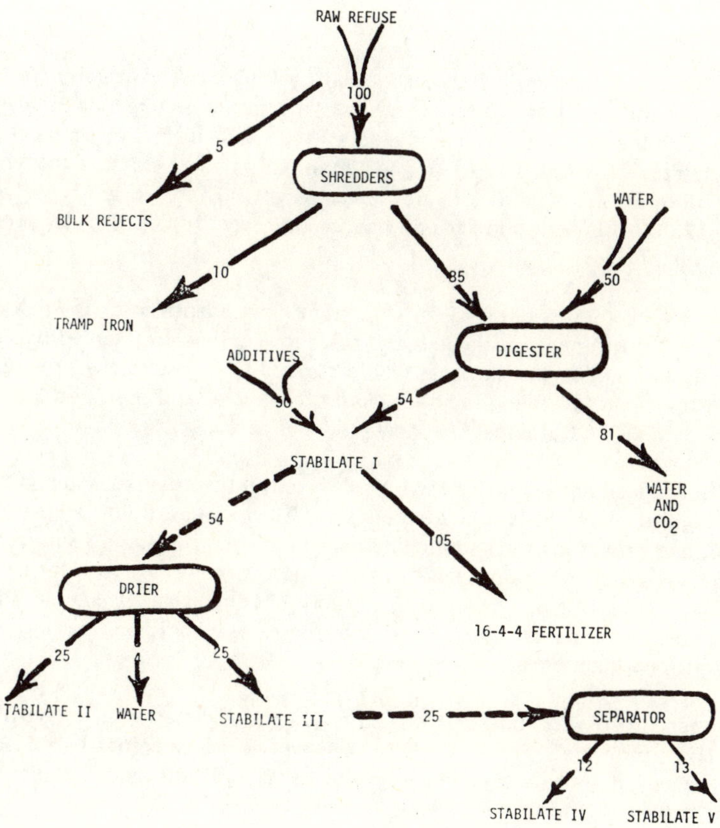

Fertilizer — The conversion process end product, Stabilate I, is basically clean, easily stored and transported. It is enriched so as to contain specified minimum percentages of nitrogen, phosphorus and potassium and qualifies under most state laws as organic in that a certain part of the nitrogen is in slow release form. The flexibility of formulation extends from unenriched compost (less than 1-1-1) to highly enriched organic fertilizer with nitrogen units covering the entire range of the intermediate formulae. The average fertilizer product bulk density is 35 pounds per cubic foot.

Tests performed at the Department of Agronomy, of Penn State University (Waddington, Duich and Moberg, Progress Report 327, *Evaluation of Fertilizer Containing Composted Refuse,* Penn State University, May 1972), indicate that this fertilizer compares favorably with Scott's Turf Builder and Borden's as a lawn fertilizer. Three Ecology formulae were tested and found to produce no appreciable fertilizer burn and are very comparable in color and growth (measured in clipping weights).

Fuel — The conversion process as presently configured produces five classifications of Stabilate. The characteristics of these classifications are shown in Table 9.1. Only the designation of Stabilate III has been proposed for use as fuel.

On the basis of calorimetric tests performed by New York University for Ecology, Inc., Stabilate III as produced by the plant has an average heating value between 7,000 and 7,500 Btu per pound. The raw municipal refuse as received has an average of 4,700 Btu per pound. Stabilate retains ⅔ of the total heating value of the raw incoming refuse and ½ its weight. This compares with approximately 12,000 Btu per pound for anthracite coal, and 18,000 Btu per pound for fuel oil.

Stabilate III's combustion characteristics seem most adaptable to the market comprising institutional and large scale residential boilers now served by coal, since only at such smaller furnaces would storability be worth the processing costs. Storage and handling facilities would be similar, and pollution abatement problems would be less expensive, as would the cost per Btu.

The effluent loadings would be low in sulfur and most other gaseous pollutants, but as high or higher in particulate content. Ash residue would be greatly increased, and provisions for its handling may be difficult in some facilities. Stabilate III's utility for large power plants and other burners with high performance fuel requirements would be notably less than that of coal and oil. It is not clear that its advantage over shredded refuse would even warrant investigation, considering processing costs.

It is important to note, however, that there has been no attempt to burn Stabilate III outside the laboratory. Trial runs will certainly be needed to determine suspension and ash characteristics, handling problems, and the adaptability of coal storage and feed mechanisms.

The value in terms of heat of this Stabilate III is estimated at $14 per ton delivered. This revenue would have to cover any portion of the conversion process not covered by the disposal fee, yet permit the user to meet the cost of transportation to the furnace, plus any storage fee at strategically located storage areas within the city, plus the cost of conversion of coal storage hoppers to larger compost storage and larger ash storage, and the cost at each facility of converting so that Stabilate III could replace coal or oil or whatever is now being used.

The cost of emission control equipment which would have to be upgraded or adjusted to handle particulate emissions anticipated would have to be weighed against the cost of upgrading some or most of these sites to conform with sulfur emissions control requirements. An engineer with the New York City Environmental Protection Administration has calculated that this usage of the refuse is competitive with the usage of raw refuse directly as a fuel in smaller facilities.

Construction Materials – Ecology, Inc. has investigated the use of Stabilate as aggregates for such products as fiberboard, cardboard and paper. The total annual demand for which Stabilate is competitive in paperboard and construction materials is estimated to be 4,343,000 tons in New York City, Hartford and Philadelphia. If it is felt that 15% of the market is within the reach of these plants, then five of them would be able to handle the market in these three cities altogether.

RELATIONSHIP OF PLANT LOCATION TO PRODUCT DISPOSAL

The facility's compatibility with heavy industry land uses is an important factor in determining locational acceptability. Unlike many traditional forms of disposal, these resource recovery facilities do not require a disproportionately large land area, they have no visible effluent, and they are not generally obnoxious neighbors in land zoned for heavy industry. This would permit inner city locations, achieving a competitive advantage in terms of haulage and transportation fees over those types of disposal facilities which require larger area.

The design of the facility and its modular composition do not permit great economies of scale beyond one thousand tons per day capacity, so that dispersion of these plants is, at this point in the investigations, feasible. This dispersed municipal configuration in an intensely developed megalopolis will permit the most flexible provision of disposal services; this process may be incorporated with others in a city's system.

Since characteristics of a 1,000 ton a day plant are nearly indistinguishable in a heavy industry land use pattern, it is superficially evident that much of the 26,000 tons a day disposal capacity of New York City or that of any major metropolitan area with a fair distribution of industrial zoning would lend itself to relief by this approach. Those metropolitan areas whose spatial distribution patterns are less adaptable for these purposes could discover through analysis some lesser extent of dependence upon the method, integrating it into their existing system, while retaining alternative disposal techniques where existing land use is incompatible or when markets are not identifiable.

As shown in Table 9.1, the true specific gravity is greater than one. This could be significant if marine disposal were contemplated, but obviously the material would need some form of restraining cover since many particles (e.g., styrofoam) would not sink. It should not be anticipated that any appreciable quantities of Stabilate will require disposal, however, since there are cheaper methods of pretreating refuse for disposal. While no attempt has been made to provide sludge disposal at this facility, the prospects for including sludge handling appear favorable at future facilities.

PROCESS TECHNOLOGY

General Description

The raw refuse received and processed at the plant is collected by the city for incineration. The routes served are primarily the residential routes in the areas adjacent to the plant. No unusual attempt is made to exclude bulky items, but rather anything that is left on the street for collection which the collectors are willing to pick up and put into the hopper of a compaction truck is brought to the facility. The city has chosen to hold itself responsible for retrieving and disposing of whatever miscellaneous bulky wastes are not marketable (usually 5% by weight).

The tramp iron (usually 10%) is removed by a dealer without reimbursement. Hopper dimensions and other hardware constraints at the facility limit the entry size of bulky waste receipts to a maximum dimension of 3' x 2' x 1'. In the facility the truck loads are dumped on to a platform, and the refuse is pushed into the hoppers using a small front end loader. This arrangement permits extraction of refrigerators, large rugs, bowling balls, or other items which may damage the hopper and the shredding mechanism, but any preferred system could be adopted for use compatible with the process.

The following is a description of the apparatus contained in the facility, which is designed so that it may be increased in capacity through modular addition of standard size units. Essentially there are three basic elements to the process, the first being the shredding segment, the second being the digester or the bacteriological segment and the third being the product finishing portion of the plant.

The first and third segments are designed for eight hour operation per day, the middle stage, of course, is designed for a twenty-four hour operation. The first segment is composed of the above-mentioned feeder process, with the dumping platform and front end loader, and two parallel lines, each with a hopper and two shredders in series connected by conveyors and magnetic separation devices. The second segment, the digester apparatus, consists of superposed, stationary decks on which the composting material is turned over by mechanical rakes which move it along the deck horizontally where it drops to the next lower deck.

The height of the composting layer on the decks is maintained at an average of 12 inches; the decks are composed of a baisc unit of eight stacked decks ten feet long and twelve feet wide. Each modular unit holds 14 tons for a design maximum retention time of 72 hours. The minimum effective length of a digester, determined by through-put/cost ratio, is eight modules, or 80 feet. Structurally it is considered to be uneconomical to go beyond 200 feet. The third segment is devoted to the manufacture of finished organic fertilizer; it consists of a screen, a dryer, compost storage, a mixer, a compactor, a granulator, a product screen, and a product storage and bagging area.

The plant is situated on a five acre site. Thirty percent of the property is devoted to vehicular access, twenty percent to storage of the product and fifty percent to the building which houses the process. Of the 42,000 square feet of building on the site, ten percent is administrative, twenty-five percent is in the shredder segment, forty percent is in the digesting segment and twenty-five percent is in the production of fertilizer. Intentionally, there is no unprocessed refuse holding capacity other than the shredder hopper, and the apron upon which the vehicles dump.

There is holding capacity provided in two 18 cubic yard open top metal boxes for separated bulky and ferrous wastes, and another for digester residue which has been removed by screening and is hauled away to disposal. The company's intentions are to phase out the use of this latter need for disposal by regrinding the screeing for inclusion in the Stabilate.

These screenings, however, are largely made up of glass, plastic, and nonferrous metals which do not enhance the quality of the Stabilate for fuel or soil conditioning uses. The purpose, however, is that eventually there will be absolutely no residue which is not processed for subsequent sale or reuse.

At one point, the Department of Public Health of the City of New York required placing of rodenticide and other rodent control procedures on a monthly basis; this was discontinued after a period of several months in which no rodents were observed. Operating personnel have noted rats on the top (undigested) deck, however. Casual site inspection does not indicate any inordinate amount of noise or odors or other displeasant emissions from the operation in its present location in an industrial area on the waterfront of the East River.

Shredding Segment

Access to this particular facility is awkward and not clearly marked. Trucks entering pass over an automatic truck scale where weight, tickets and billing paper work is handled, and then up a ramp around to a charging platform.

Queuing space is confined; the ramp will allow only one truck at a time. Two parallel feeders, 60 feet by 8 feet and two apron conveyors 30 feet by 6 feet carry the wastes past two manual sorting stations to the primary shredders. Manual separation is designed to remove large chunks of metal, mattresses, springs, rugs, and other hard to handle items which could damage or bind the machinery; the purpose of removal is not for salvage, the wastes removed (approximately 5% by weight) are simply stored for Sanitation Department removal.

Approximately 10% of the refuse is magnetically separated for removal by a broker at no fee. The two primary shredders, arranged in parallel, each have 250 horsepower motors. They shred the refuse so that about 85% is ½ inch or less according to tests performed by the company, increasing the incoming refuse bulk density from 15 pounds per cubic foot to about 25 pounds per cubic foot. Particle size determinations are run on 5 pound grab samples of the product of both sets of shredders and grate adjustments can be made to achieve the optimal particle size for composting.

Conveyors transport the shredded refuse across magnetized head pulleys which remove 90 to 95% of the tramp iron, and discharge the remaining material to the secondary shredders. These shredders further reduce the particle size to the point where approximately 90% is ⅜ inch or less, further increasing bulk density to about 40 pounds per cubic foot according to the company's estimation.

Both pairs of shredders, similarly constructed and operated, have a single horizontal rotor shaft with swinging hammers. The design rate of all four shredders is 20 tons per hour. Each is powered by a 250-horsepower, open, drip-proof, electric motor that powers the rotor shaft by means of a flexible coupling.

Refuse is crushed between the swinging hammers and the grates. The refuse is pounded until it is small enough to pass through the grates, located below the rotor shaft. The grates are adjusted after every 200 tons so that they almost touch the swinging hammers. These adjustments tend to improve the efficiency of the shredders in reducing the particle size.

The grate openings in the primary shredder are 6" x 8". The grates of the secondary shredder, have openings that are 1" x 6". The secondary shredder grates were extended from 135° to 180° when it was found that no bulky bypass was occurring. Each shredder is provided with a cover for access to the hammers and grates.

Digestion Segment

The shredded refuse which contains about 30 to 40% moisture is then carried by conveyor to a storage tank, and then to the top deck of the digestor. A chain loop arrangement provides drive for a series of parallel bars which tow rakes through toe compost pile. Four rakes are spaced about 24" apart on each bar. These bars are about 20 to 25 feet apart on the chain and travel on one chain southbound on one platform, drop down to a lower platform and travel the return northbound trip. The normal operating speed of the rakes is between 30 and 45 feet per minute. These rakes provide the only means of physical agitation that the composting material receives in the digester, both for forward motion and for aeration and mixing. The spacing between the furrows on a bar is such that consecutive furrows travel in slightly offset paths, providing thorough agitation in this continuous flow process. The process provides considerable flexibility for independent control and measurement of the determining factors of composting. Some of the techniques for providing these controls are proprietary, but a general description follows:

(a) The supply of oxygen is maintained by the forced air heating system; the rate of aeration is then determined by the particle size, the height of the pile and the agitation. The bed agitation is controlled separately for each pair of decks by the speed and the position adjustment of the raking apparatus (see (f) below). While the control of neither supply nor contact is completely independent, the range of operating requirements is such that both may be adequately achieved without disrupting the other factors such as detention time or the air temperature.

(b) Particle size may be optimized by adjustment of primary and secondary shredder grates. Experimentaion has determined optimal adjustment for given agitation, temperature and moisture ranges.

(c) Moisture-recording sensors are set in probes in each of the decks to measure moisture content of the compost pile. Water may be sprayed at the beginning of each deck as needed for increasing moisture. There is no provision for downward adjustment other than stopping the spray and increasing the air temperature. The optimal moisture content on the top deck is 50 to 55%.

(d) pH is measured by the same deck-mounted probes. It can be controlled by the addition of buffering agents to the moisturizing spray, separately for each deck. pH is thus not independent of moisture controls in that it may not be controlled when no moisture is needed, but in routine operation that is not a problem. Initially, as microbial action commences, pH will drop quickly due to carbohydrate breakdown into organic acids. After the consumption of the acids, the pH tends to rise to 7.5 but seldom higher.

(e) Temperature can be controlled by an air heating system separately adjustable for each deck. Probes are used to measure air temperature and pile temperature. While air temperature can be controlled directly, pile temperature adjustments can only be influenced through air temperature or other controls such as addition of hot water or agitation. The air heating system is presently adjusted to maintain ambient air temperature about 10°F above pile temperature on the top six decks and at 150°F on the bottom two decks.

(f) The speed that the pile travels across the decks can be adjusted separately for each pair of decks, since a single chain drives the rakes for two decks. Leveling bars which travel in front of each rake assembly can be adjusted between a minimum pulp layer height of 8" and a maximum of 16" Chain speeds can be adjusted between 3 feet per minute minimum and 45 feet per minute maximum. The rake delivery can be measured with automatic impact sensors which feed back instructions to the drive chain. These may be adjusted by overrides which provide for temperature control or mixing or the other parameters.

Table 9.2 following indicates the optimal parameter settings. The detention time is normally adjusted to 40 hours.

TABLE 9.2: PARAMETER SETTINGS

Deck	pH	% Moisture	Pile Temperature °F	Air Temperature °F
I	5.5-6.2	50-55	90-120	125-130
II	5.5-6.2	45-50	130	135-140
III	5.5-6.2	40-45	120	125-130
IV	5.7-6.4	35-40	115	120-125
V	5.9-6.6	30-35	110	115-120
VI	6.1-6.8	25-30	110	115-120
VII	6.8-7.0	15-20	90-100	150
VIII	7.0	10-15	90-100	150

The Warburg respirometer test, similar to BOD tests, is used to determine the degree of degradation of available carbon. The company states that these tests indicate no adverse effect due to normal fluctuation in paper content.

Much of the system is automated. Further automation provides safety control devices such as flow sensors and, oxygen sensors, excessive binding or torque demands, as well as fire and system component breakdown alarms.

Product Segment

The third segment, the product treatment apparatus, is arranged for fertilizer production. It is described as follows:

The compost screen (Number 4 mesh) is fed by conveyors from the bottom of the digester. It removes approximately

10% of the oversized material for regrinding, mostly undecomposed substances, including plastics and nonferrous metals. This is followed by a compost storage tank, which holds some biologically stable material in stockpile for process flow equilibrium. Several compartments may be required depending upon chemical consistency needed in the various product specifications. This is followed by a mixer, fed by a conveyor belt, through which solid additives are incorporated.

The next process uses a compactor, granulator and product screen. They are adjustable to achieve various particle size and other physical characteristics in the product. It is followed by another storage tank which feeds the bagging operation.

Automatic additives control is also provided in the production of fertilizer for the nitrogen enrichment. The final N-P-K enrichment is accomplished by the addition of solid additives after the composting process. Quality control sensors are used for gross indicators of carbon and nitrogen in the raw material, and in the composting pulp.

ASSESSMENT

Process and Technology

The technology involved seems to be fairly well developed. The stability of the product has not been rigorously determined. New York City engineers have reported faint odors while experimenting with the product submerged in sealed containers of sea water, suggesting possible degradation. Strobel and Rongved, with a similar product, has proposed sterilization by irradiation, indicating that they doubt the effectiveness of the pasteurizing effect of the composting process.

The health effects of the use of compost for edible crops has not been well researched at present. The company is continuing to investigate product characteristics, of course, since they affect all of the products' usability. Otherwise, the general technology of composting is commonly considered well developed. This specific process is still in developmental stages, and should be considered as such until successful operation on a scale of 1,000 tpd.

Environmental Influence

Assessment of the influence on the environment of proliferation of this concept must compare the impact of the system with the systems it would replace not only at the point of processing or disposal but also at the point of intended usage of the products. No lasting environmental degradation was observed at the facility from either the conversion process or the subsequent processing. Noise, configuration, and potential odors confine potential locational choices for esthetic reasons, but properly located, environmental degradation appears negligible. The impact of this system, then, would be positive to the extent that it relieved the need for usage of alternative polluting methods.

Solid outflow from the plant is in four categories: bulky rejects, ferrous scrap, noncompostable residue, and Stabilate. The bulky, such as appliances, rugs, etc., constitute 5% of the waste stream received and are the only major fraction of the incoming stream not intended for reuse. It is extracted without alteration and is removed by the city for land disposal; it is typically nonputrescible and can be mixed with demolition wastes or other clean fill.

The ferrous scrap, 10% by weight of the incoming, is shredded and extracted for removal by scrap dealers. Presumably, most of this material is recycled. The residue comprises plastics, nonferrous metals, glass fragments and other particles which have passed through shredding and digesting segments of the process and are subsequently removed by screening. This fraction is being reshredded and/or reprocessed for inclusion in the Stabilate with minimal loss of quality.

The success is of course dependent upon degree of reprocessing and on the subsequent use of the Stabilate. Stabilate is wholly intended for subsequent usage. To the extent that its use is not demanded at any price, it may be satisfactorily landfilled. Its characteristics in landfill are only speculative at this point. There is presently no removal of heavy metals or of many of the components of leachate except for ferrous extraction and pasteurization of most pathogens.

Some of the potential for gas production has been exhausted, but some doubt remains that on a volumetric basis, in place, there would be significant improvement, of course, on a basis of refuse handled. It was estimated by an engineer with the New York City Environmental Protection Administration that only ⅓ the volume of the raw refuse landfill would be required, not including any reduced cover requirements. Carbon dioxide, methane, water vapor and other gaseous emissions will occur, of course, but testing is needed to determine the degree of problem, if any.

Subsequent usage of Stabilate as a component of fertilizer adds to the soil some trace amounts of whatever miscellaneous elements were present in the waste stream. Considerable mixing occurs during the handling process, however, and the small dosages applied in fertilization of soil make investigation of resulting environmental damage seem futile.

Utilization as fuel produces a potential air pollutant which must be weighed against usage of alternative fuels. Particulate emissions will be high. The presence of heavy metals, plastics or whatever chemicals comprised in that day's collection could become the component of the day's emissions.

Experience in monitoring existing incinerators' emissions will be useful in determining the environmental impact from specific solid waste streams. It is not anticipated that this impact will be considerable.

The nature of the product of the construction materials is such that little human or environmental contact will occur during usage or upon discard. Other miscellaneous uses, such as for kitty litter, appear likewise to offer no particular pollutant threat.

ENERGY FROM SOLID WASTE 1974

by Frederick R. Jackson

Pollution Technology Review No. 8

Energy Technology Review No. 1

The solid waste disposal problem is reaching alarming proportions everywhere. The United States alone produces close to 300 million tons of solid waste per year, which is equivalent to about one ton per person.

At the present time the prevalent methods of disposal are dumping and sanitary landfill. Municipal incineration disposes of a small portion only, with attendant high capital and operating costs.

Many methods have been proposed for coping with the problem, such as source separation, source reduction, or material recovery. However, with the energy crisis descending upon us, producing energy from waste is becoming more and more attractive. An estimate currently making the rounds in financial circles is that when the price of crude oil reaches $7.00 a barrel, alternate sources of energy become practicable.

This book is based primarily upon information from studies conducted under the auspices of the EPA. Its foremost topic is burning of solid wastes to create steam directly. The air pollution problem created by burning can be solved quite easily with known technology. Solid wastes are low in sulfur, consequently there is no SO_2 removal problem.

Another technique that may assume more importance in the future is the controlled pyrolysis of wastes, yielding so-called pyrolysis gas or oil. Chapter seven is devoted to this. The final chapter discusses European practice, which is historically far more extensive than that of the U.S. A condensed table of contents follows here:

ISBN 0-8155-0528-0

163 pages

SANITARY LANDFILL TECHNOLOGY 1974

by Samuel Weiss

Pollution Technology Review No. 10

This Pollution Technology Control Review surveys all the latest technical information on this subject. It is based primarily on studies conducted by industrial or engineering firms, under the auspices of the Environmental Protection Agency.

Sanitary landfill practice is an engineering method of disposing of solid wastes on land, whether they are of municipal or industrial origin. There is no resemblance to the old-fashioned garbage and rubbish dump. There are no fires, no obnoxious fumes or smoke, no flies and no rodents or other scavengers.

As stated in the introduction to this book, a sanitary landfill is not only an acceptable and economic method of solid waste disposal, it provides also an excellent way to improve the commercial value of otherwise unsuitable or marginal land areas within a few years.

A partial and condensed table of contents follows here.

ISBN 0-8155-0542-6 **300 pages**

POLLUTION DETECTION AND MONITORING HANDBOOK 1974

by Marshall Sittig

Environmental Technology Handbook No. 1

This handbook contains methods for the detection and monitoring of pollutants in industrial effluents, notably air and water.

Such methods are prerequisites for any kind of management of environmental quality. Of equal importance are the data handling systems that must be used to collect and interpret the measurements, regardless of whether manual or automated identifying and monitoring techniques are chosen.

Generally the most important single factor in determining whether manual or automated methods of analysis and level monitoring should be used, is the required frequency interval. In the manual approach costs vary almost directly with the measurement frequency. At monitoring intervals oftener than once a day, or when a continuous watch must be maintained, installation of automatic sensing is more economical, provided reliable automatic sensors are available.

For meaningful control and prevention of pollution the accepted standards and parameters for air and water quality must be known, whether they have been legislated or not. Acceptable levels for each type of impurity from methylmercury to dissolved chlorides or carbon particles must be known by the industrialist or public health official, and this book, with its 1633 references makes a massive attempt to communicate such levels.

As a guideline for the industrial user, data on the toxicity of the various pollutants are included. Such data on the deleterious effects were taken from various government publications. This book is not a treatise of pharmacology or industrial toxicology, but a guide to the accurate evaluation of pollutant levels in a given effluent or ambient substrate.

Whenever applicable, the name of the pollutant is followed by chemical and other information. After this come directions for sampling in air or water or both, as the case may be. Thereafter are given qualitative and quantitative analytical methods suitable for identification and measurement of the degree or level of pollution. This comprehensive information is followed by measurement techniques suitable for repeated or continuous monitoring. Preference is given to methods giving reproducible results in environmental quality control, as recommended by the EPA and other U.S. government agencies. This is followed in each case by many up-to-date references to government publications, patents, and journal articles.

Arrangement is encyclopedic. The book is thus a worthy companion to the author's POLLUTANT REMOVAL HANDBOOK available from the same publisher.

Pollution control is a serious business, and in the design of a monitoring program the specific objectives to be served must be clearly in mind from the very beginning. Otherwise the effort most probably will not result in an efficient and meaningful program. The length of the initial survey, selection of monitoring points, parameter coverage, and sampling frequency all depend on the specific objectives and the timeframe in which the objectives must be achieved. These variables also have a very significant impact on monitoring costs.

Monitoring for the purpose of long-term trend identification, evaluation of standards compliance, and total management of environmental quality must, of course, be a continuing program without end.

It is hoped sincerely that this book, together with its companion volume POLLUTANT REMOVAL HANDBOOK, will do much to establish and sustain such antipollution programs.

A partial and condensed table of contents follows here in which subheadings are indicated below for only the first few entries. The book contains a total of 88 subject entries arranged in an alphabetical and encyclopedic fashion. The subject name refers to the polluting substance, and the text underneath each entry tells how to measure and monitor pollution by said substance:

POLLUTION INSTRUMENTATION
Size of the Market
References

SAMPLING
Air and Gas Sampling
Water Sampling
References

SUBSTANCES AND MATERIALS

ACIDS AND ALKALIS

ALDEHYDES

ALUMINUM

AMMONIA

ANTIMONY

ARSENIC

ASBESTOS

AUTOMOTIVE EXHAUST

BACTERIA

BARIUM

BERYLLIUM

BISMUTH

BORON

BROMIDES

CADMIUM

CALCIUM

CARBOHYDRATES

CARBON

CARBON DIOXIDE

CARBON MONOXIDE

CHLORIDES

CHLORINATED HYDROCARBONS

CHLORINE

CHROMIUM

COBALT

COPPER

CYANIDES

ETHYLENE

FLUORIDES

HALOGENS

HALOGEN COMPOUNDS

HYDROCARBONS

HYDROGEN CHLORIDE

HYDROGEN SULFIDE

IODIDES

IRON

LEAD

LEAD ALKYLS

MAGNESIUM

MANGANESE

MERCAPTANS

MERCURY

MOLYBDENUM

NICKEL

NICKEL CARBONYL

NITRATES

NITRIC ACID

NITRITES

NITROGEN COMPOUNDS

NITROGEN OXIDES

ODOROUS COMPOUNDS
Measurement in Air
Sampling Methods
Qualitative Analyses
Quantitative Methods (Organoleptic)
Quantitative Methods (Instrumental)
Measurement in Water
References

OIL AND GREASE

ORGANICS

OXYGEN DEMAND

OZONE AND OXIDANTS

PARTICULATES

PEROXYACETYL NITRATE (PAN)

PESTICIDES

PHENOLS

PHOSGENE

PHOSPHATES

PHOSPHITES

PHOSPHORIC ACID

PHOSPHORUS

PHOSPHORUS COMPOUNDS (ORG.)

POLLEN

POLYNUCLEAR AROMATICS

POTASSIUM

RADIOACTIVE MATERIALS

SEDIMENTS

SELENIUM

SILVER

SODIUM

STRONTIUM

SULFATES

SULFIDES

SULFITES

SULFUR COMPOUNDS (ORG.)

SULFUR DIOXIDE

SULFUR TRIOXIDE

SULFURIC ACID

SURFACTANTS

TIN

TITANIUM

URANIUM

VANADIUM

ZINC

ZIRCONIUM

FUTURE TRENDS

This book should prove to be very useful as an overall reference volume. The organization is clear, and the text is especially well arranged. The tables and figures are extensive and will be of value to students, engineers, and pollution control authorities.

ISBN 0-8155-0529-9

401 pages

POLLUTION CONTROL IN THE ORGANIC CHEMICAL INDUSTRY 1974

by Marshall Sittig

Pollution Technology Review No. 9

Wastes from plants manufacturing identical compounds may be quite dissimilar, because of the difference in processes and raw materials. Since the organic chemical industry is now largely petrochemical, preference has been given to waste treatment from such operations.

Detailed treatability evaluation of each waste stream has become a prerequisite for any profitable product.

Physical methods for waste and by-product removal include gravity separation, air flotation, filtration, evaporation and adsorption techniques.

Chemical methods include precipitation, polyelectrolyte treatment, oxidation and other chemical conditioning such as neutralization.

Biological treatment methods comprise activated sludge and its modifications, trickling filters, aerated lagoons, and waste stabilization ponds.

This book intends to assist the chemical engineer in treating chemical waste products and effluents in conjunction with prudent raw material choices and thus bring about process economics. It presents condensed vital data from government and other sources of information that are scattered and difficult to pull together.

A partial and condensed table of contents follows here.

Dinitrotoluene from Nitrotoluene

Disulfoton from Ethanol

Dyes and Pigments from Aromatics

Epichlorohydrin from Allyl Chloride

Ethylbenzene from Benzene

Ethyl Chloride by
Hydrochlorination of Ethylene
Chlorination of Ethane
Hydrochlorination of Ethanol

Ethylene or Propylene from Paraffins

Ethylene Dichloride from Ethylene

Ethylene Glycol from Ethylene Oxide

Ethylene Oxide from Ethylene

Formaldehyde from Methanol

Long Chain Alcohols from Ethylene Oligomers

Malathion® from Methanol

Methanol from Natural Gas

Methylamines from Methanol + Ammonia

Methyl Bromide from Methanol

Methyl Methacrylate from Acetone Cyanohydrin

Nitrobenzene from Benzene

Nitrochlorobenzene from Chlorobenzene

Nitroparaffins from Paraffins

Oxo Products from Olefins

Parathion from p-Nitrophenol

Perchloroethylene from Propane

Phenol from Chlorobenzene

Phenol from Cumene

Phorate from Ethanol

Phosgene from CO + Cl_2

Phthalic Anhydride from Naphthalene or Xylene

Propylene Glycol from Propylene Oxide

Propylene Oxide from Propylene

Styrene from Ethylbenzene

SNG from Crude Oil

2,4,5-T from Trichlorophenol

Terephthalic Acid from Xylene

Tetraethyl Lead from Ethyl Chloride

Trichloroethane from Vinyl Chloride

Trichloroethylene by
Chlorination, then by
Dehydrochlorination of Acetylene
or by Oxyhydrochlorination
of Dichloroethane

Trifluralin from Perfluoromethylchlorobenzene

Vinyl Acetate from Ethylene and Acetic Acid

Vinyl Chloride from Acetylene

Vinyl Chloride from Ethylene Dichloride

AIR POLLUTION CONTROL
Halogen Acids
Halogenated Hydrocarbons
Halogens
Hydrocarbons
Particulates
Sulfur Compounds
Coking of Coal

WATER POLLUTION CONTROL
Primary
Oil Separation
Equalization
Neutralization
Sedimentation
Flotation
Flocculation
Nutrient Addition
Secondary
Activated Sludge
Extended Aeration
Trickling Filters
Aerated Lagoons
Waste Stabilization Ponds
Chemical Oxidation
Nitrification—Denitrification

Tertiary
Chemical Precipitation
Gas Stripping
Microstraining
Carbon Adsorption
Electrodialysis
Ion Exchange
Evaporation
Reverse Osmosis
Chlorination
Rapid Sand Filtration
Sludge Handling
Aerobic Digestion
Anaerobic Digestion
Wet Oxidation
Thickening
Lagooning
Sand Drying Beds
Vacuum Filtration
Filtration
Land Disposal
Incineration
Sea Disposal
Ultimate Disposal
Thermal Oxidation
Deep Well Disposal

OVERALL WATER POLLUTION CONTROL MODELS
BPCTCA: Best Practicable Control Technology Currently Available
BATEA: Best Available Technology Economically Available
BADCT: Best Available Demonstrated Control Technology

THE ECONOMICS OF POLLUTION CONTROL

FUTURE TRENDS

ISBN 0-8155-0536-1

305 pages